Enzyme-Mediated Stereoselective Synthesis

Enzyme-Mediated Stereoselective Synthesis

Special Issue Editor

Stefano Serra

MDPI • Basel • Beijing • Wuhan • Barcelona • Belgrade

MDPI

Special Issue Editor
Stefano Serra
Istituto di Chimica del Riconoscimento
Molecolare (ICRM)
Italy

Editorial Office
MDPI
St. Alban-Anlage 66
4052 Basel, Switzerland

This is a reprint of articles from the Special Issue published online in the open access journal *Catalysts* (ISSN 2073-4344) from 2018 to 2019 (available at: https://www.mdpi.com/journal/catalysts/special issues/stereoselective synthesis).

For citation purposes, cite each article independently as indicated on the article page online and as indicated below:

LastName, A.A.; LastName, B.B.; LastName, C.C. Article Title. *Journal Name* **Year**, *Article Number*, Page Range.

ISBN 978-3-03921-936-0 (Pbk)
ISBN 978-3-03921-937-7 (PDF)

Contents

About the Special Issue Editor

Stefano Serra is a Researcher at the National Research Council (CNR) in the Istituto di Chimica del Riconoscimento Molecolare (ICRM), Milano, Italy. He received his Laurea from the University of Pavia in 1995, where he was working on the synthesis of the antitumoral triterpene saponaceolide B. One year later, he joined the group of Professor Fuganti at the Politecnico of Milano, where he carried out studies on the use of enzymes in organic synthesis and on the stereoselective preparation of flavor and fragrance molecules. In 2000, he was awarded a PhD in Industrial Chemistry from the University of Milano, and in 2001, he was a Researcher of the National Research Council. His scientific activity has been devoted to many aspects of organic synthesis and, above all, to the enantioselective preparation of chiral compounds and to the development of new synthetic methods.

catalysts

Editorial

Enzyme-Mediated Stereoselective Synthesis

Stefano Serra[ID]

C.N.R. Istituto di Chimica del Riconoscimento Molecolare, Via Mancinelli 7, 20131 Milano, Italy;
stefano.serra@cnr.it or stefano.serra@polimi.it; Tel.: +39-02-2399-3076

Received: 6 August 2019; Accepted: 8 August 2019; Published: 11 September 2019

All of us know very well the importance of the catalysis in organic synthesis. Catalyzed reactions are usually preferred when planning a new synthetic approach and the choice of the proper catalyst is of pivotal relevance.

In this context, the use of biocatalysts in organic synthesis has grown steadily during the last fifty years. Presently, chemists have become accustomed to the idea that the use of enzymes is essential in a modern synthetic laboratory. It does not matter if a given transformation is performed using whole cell microorganisms or using an isolated enzyme. In both cases, the specific activity and selectivity of one or more enzymes is exploited. In fact, biocatalysts allow for performing a number of chemical reactions with high regio- and stereoselectivity and a large number of biocatalyzed industrial processes have already been established.

This Special Issue has been planned in order to collect studies focused on the exploitation of enzyme stereoselectivity for the synthesis of relevant chemicals, such as innovative materials, active pharmaceutical ingredients, natural products, flavours and fragrances, and any other kind of bioactive compounds.

Overall, the Special Issue has gathered six articles and one review.

A first paper from Serra and De Simeis [1] describes the stereoselective synthesis of the enantiomeric forms of the alcohol (2,6,6-trimethyltetrahydro-2*H*-pyran-2-yl)methanol, which are potential chiral building blocks for the preparation of different natural terpenes. Two different catalytic approaches are reported. The first is based on the stereospecific (+)-10-camphorsulfonic acid catalyzed cyclization of 2-methyl-5-(2-methyloxiran-2-yl)pentan-2-ol enantiomers, which are synthesizable from linalool enantiomers, easily available from the chiral pool. The second synthetic approach is based on the lipase-mediated resolution of the target tetrahydropyranyl alcohol, which is available in racemic form starting from the industrial intermediate, dehydrolinalool. In this work, a large-scale resolution procedure that exploits the opposite enantioselectivity of Novozym® 435 lipase and lipase AK in the acetylation reaction of (2,6,6-trimethyltetrahydro-2*H*-pyran-2-yl)methanol is described. The two enantiomeric forms of the latter alcohol were then employed for the first stereoselective synthesis of both enantiomers of the flavor linaloyl oxide.

Concerning resolution based processes, Boratyński et al. [2] reported a novel microbial approach to the production of enantiomerically enriched aroma compounds based on solid-state fermentation using agroindustrial side-stream feedstocks. Twenty-five filamentous fungi were screened for lipase activity and enantioselective hydrolysis of the racemic flavours 1-phenylethyl acetate, *trans* and *cis* whisky lactones, γ-decalactone, δ-decalactone, and *cis*-3a,4,7,7a-tetrahydro-1(3*H*)-isobenzofuranone. Solid-state fermentation was conducted with linseed and rapeseed cakes and the results highlight the potential economic value of solid-state fermentation as an alternative to more expensive processes conducted in submerged fermentation.

Two relevant works exploit the potential of biocatalysed reduction reactions. A first contribution from Brenna et al. [3] describes the enantioselective synthesis of β-nitroalcohols by enzyme-mediated reduction of α-nitroketones. In this work, the use of commercial alcohol dehydrogenases (ADHs) for the reduction of aromatic and aliphatic nitroketones is investigated. High conversions and enantioselectivities were achieved with two specific ADHs, affording either the (*S*) or (*R*)-enantiomer of the corresponding nitroalcohols. The further manipulation of the enantioenriched nitroalcohols into protected aminoalcohols is also described. A second contribution from Gotor-Fernández et al. [4] reports the preparation of chiral 1,4-diaryl-1,4-diols through ADH-catalyzed bioreduction of the corresponding bulky 1,4-diaryl-1,4-diketones. Among the different enzymatic preparations used, ADH from *Ralstonia* sp. (RasADH) overexpressed in *E. coli* afforded the best results in terms of conversions and diastereo- and enantiomeric excess, allowing the preparation of a set of 1,4-diaryl-1,4-diols bearing different pattern substitutions in the aromatic ring.

Concerning enantioenriched alcohol derivatives, Serra and De Simeis [5] synthesized natural hydroxy fatty acids by means of a biocatalytic hydration reaction of natural fatty acids. The study describes the use of the probiotic bacterium *Lactobacillus rhamnosus* (ATCC 53103) as whole-cell biocatalyst for the hydration of the most common unsaturated octadecanoic acids, namely oleic acid, linoleic acid, and linolenic acid. The reaction proceeds with high regio- and stereoselectivity. Only 10-hydroxy derivatives were formed and the resulting (*R*)-10-hydroxystearic acid, (*S*)-(12*Z*)-10-hydroxy-octadecenoic acid, and (*S*)-(12*Z*,15*Z*)-10-hydroxy-octadecadienoic acid were obtained in very high enantiomeric purity (ee > 95%). The biotransformation protocol is stereoselective, scalable, and holds preparative significance, suitable for the production of high-value flavor ingredients.

Yan et al. [6] report a study on the use of modified pullulan polysaccharide for lipase immobilization. *Burkholderia cepacia* lipase, immobilized on this new material, showed a very good performance, significantly shortened the reaction equilibrium time in transesterification reaction, and exhibited good operational stability. It was found that the enlarged spherical surface of the particle led to high-immobilized efficiency, resulting in the improvement of enantioselectivity in chiral resolution.

Finally, Yun et al. [7] contribute with a review on the recent advances in synthetic applications of ω-transaminases for the production of chiral amines. Recent developments in protein engineering techniques and incorporation of ω-transaminases in multi-enzymatic cascades are there detailed. This work shows that ω-transaminases are efficient catalysts for the synthesis of enantiomerically pure amines.

Overall, these seven contributions give the reader fresh insights on the use of the biocatalysis for the stereoselective synthesis of some relevant chemical compounds (high value products or chiral building blocks) whose preparation, by means of the classical asymmetric processes, presents many drawbacks or cannot be performed.

Funding: This research was funded by [Cariplo Foundation], grant number [2017-1015 SOAVE (*Seed and vegetable Oils Active Valorization through Enzymes*)] and by [Regione Lombardia], grant number [228775 VIPCAT (*Value Added Innovative Protocols for Catalytic Transformations*)].

Conflicts of Interest: The author declares no conflict of interest.

References

1. Serra, S.; De Simeis, D. Two complementary synthetic approaches to the enantiomeric forms of the chiral building block (2,6,6-trimethyltetrahydro-2*H*-pyran-2-yl) methanol: Application to the stereospecific preparation of the natural flavor linaloyl oxide. *Catalysts* **2018**, *8*, 362. [CrossRef]
2. Boratyński, F.; Szczepańska, E.; Grudniewska, A.; Olejniczak, T. Microbial kinetic resolution of aroma compounds using solid-state fermentation. *Catalysts* **2018**, *8*, 28. [CrossRef]
3. Tentori, F.; Brenna, E.; Colombo, D.; Crotti, M.; Gatti, F.G.; Ghezzi, M.C.; Pedrocchi-Fantoni, G. Biocatalytic approach to chiral β-nitroalcohols by enantioselective alcohol dehydrogenase-mediated reduction of α-nitroketones. *Catalysts* **2018**, *8*, 308. [CrossRef]
4. Mourelle-Insua, Á.; De Gonzalo, G.; Lavandera, I.; Gotor-Fernández, V. Stereoselective enzymatic reduction of 1,4-diaryl-1,4-diones to the corresponding diols employing alcohol dehydrogenases. *Catalysts* **2018**, *8*, 150. [CrossRef]

5. Serra, S.; De Simeis, D. Use of *Lactobacillus rhamnosus* (ATCC 53103) as whole-cell biocatalyst for the regio-and stereoselective hydration of oleic, linoleic, and linolenic acid. *Catalysts* **2018**, *8*, 109. [CrossRef]

6. Xu, L.; Cui, G.; Ke, C.; Fan, Y.; Yan, Y. Immobilized burkholderia cepacia lipase on ph-responsive pullulan derivatives with improved enantioselectivity in chiral resolution. *Catalysts* **2018**, *8*, 13. [CrossRef]

7. Patil, M.D.; Grogan, G.; Bommarius, A.; Yun, H. Recent advances in ω-transaminase-mediated biocatalysis for the enantioselective synthesis of chiral amines. *Catalysts* **2018**, *8*, 254. [CrossRef]

catalysts

MDPI

Article

Two Complementary Synthetic Approaches to the Enantiomeric Forms of the Chiral Building Block (2,6,6-Trimethyltetrahydro-2*H*-pyran-2-yl)methanol: Application to the Stereospecific Preparation of the Natural Flavor Linaloyl Oxide

Stefano Serra * and **Davide De Simeis**

Consiglio Nazionale delle Ricerche (C.N.R.) Istituto di Chimica del Riconoscimento Molecolare, Via Mancinelli 7, 20131 Milano, Italy; dav.biotec01@gmail.com
* Correspondence: stefano.serra@cnr.it or stefano.serra@polimi.it; Tel.: +39-02-2399 3076

Received: 12 August 2018; Accepted: 26 August 2018; Published: 28 August 2018

check for
updates

Abstract: The enantiomeric forms of the alcohol (2,6,6-trimethyltetrahydro-2*H*-pyran-2-yl)methanol are potential chiral building blocks for the stereoselective synthesis of different natural terpenes. Here, we describe their preparation by means of two different synthetic approaches. The first is based on the stereospecific (+)-10-camphorsulfonic acid (CSA)-catalyzed cyclization of (*R*)- and (*S*)-2-methyl-5-(2-methyloxiran-2-yl)pentan-2-ol, which were in turn synthesized from (*R*)- and (*S*)-linalool, respectively. The latter monoterpenes are easily available from the chiral pool, with different optical purity. As our synthesis makes use of the intermediate 2,6-dimethyloct-7-ene-2,6-diol, whose enantiopurity can be improved through fractional crystallization, we obtained (2,6,6-trimethyltetrahydro-2*H*-pyran-2-yl)methanol enantiomers in an almost enantiopure form. The second synthetic approach is based on the lipase-mediated resolution of the aforementioned tetrahydropyranyl alcohol, which was prepared in racemic form starting from the industrial intermediate, dehydrolinalool. In this work, we report a large-scale resolution procedure that exploits the opposite enantioselectivity of Novozym® 435 lipase and lipase AK in the acetylation reaction of (2,6,6-trimethyltetrahydro-2*H*-pyran-2-yl)methanol. The two enantiomeric forms of the latter alcohol were employed for the first stereoselective synthesis of both enantiomers of the flavor, linaloyl oxide (2,2,6-trimethyl-6-vinyltetrahydro-2*H*-pyran).

Keywords: linalool; cyclization; enantioselective synthesis; enzyme-mediated resolution; lipases; flavors; monoterpenes; linaloyl oxide

---

1. Introduction

Tetrahydropyrane and tetrahydrofurane derivatives having the general framework of type **2** and **3** (Figure 1) are quite common naturally occurring compounds. The biosynthesis of the main part of these ethers is based on the cyclization of a terpenic alcohols of type **1**, and follows two main pathways, both involving the intramolecular nucleophilic addition of a hydroxyl functional group to a terminal prenyl group [1,2].

A first path requires the preliminary activation of the double bond through its transformation in epoxide or halonium derivatives, which easily cyclizes to give the corresponding ether derivatives bearing an additional hydroxyl or halogen functional group. Due to the general interest in this kind of chemical transformation in organic synthesis, its regio- and stereochemical outcomes were studied in depth [3–5]. In addition, the monoterpenes possessing a hydroxyl group and a vinyl group (E = OH,

R = vinyl) are well-known flavor components collectively called linalool oxides, regardless of the fact that they possess a six- or five-membered ring [6].

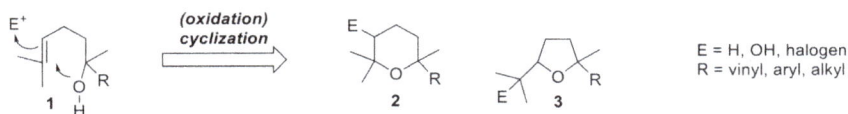

Figure 1. The transformation of terpenic alcohols of type **1** into tetrahydropyrane and tetrahydrofurane derivatives **2** and **3**, respectively.

On the contrary, the cyclization path which does not involve a preliminary oxidation reaction affords tetrahydropyrane and tetrahydrofurane derivatives devoid of a hydroxyl group (E = H). Although a number of pyranoid derivatives possessing these structural frameworks were identified in nature, only few stereoselective approaches to their synthesis are described so far. For example linaloyl oxide (Compound **4**) enantiomers (Figure 2) are relevant flavor and fragrance components [7–20], sydowic acid (Compound **5**) is a bioactive sesquiterpene produced by the mold *Aspergillus sidowii* [21], the diterpenic diol (Compound **6**) was isolated from the African plant *Anisopappus pinnatifidus* [22], and triterpene panaxadiol (Compound **7**) is a bioactive component of the ginseng extract [23]. These terpenes share in common the same tetrahydropyranyl moiety which contains a quaternary stereocenter, whose stereoselective construction is especially demanding.

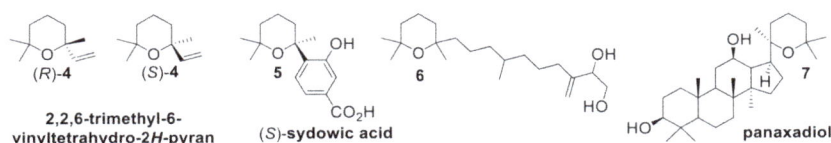

Figure 2. Representative examples of natural terpenes (mono-, sesqui-, di-, and triterpenes) bearing the 2,6,6-trimethyltetrahydro-2*H*-pyran-2-yl moiety in their molecular framework.

As we are involved in a number of studies describing the stereoselective synthesis of flavors and fragrances [24–28], and as we recently reported a new enantioselective synthesis of pyranoid linalool oxide isomers [29], we became interested in expanding our studies to the above-described class of compounds. More specifically, we envisaged that the enantiomeric forms of the primary alcohol (2,6,6-trimethyltetrahydro-2*H*-pyran-2-yl)methanol (Compound **8**; Figure 3) could be regarded as potential building blocks for the stereoselective synthesis of this kind of natural product.

According to our retrosynthetic analysis, we devised two different synthetic methods for the stereoselective preparation of alcohol **8**. Both approaches are based on an enantioselective-catalyzed reaction as key step. The first procedure takes advantage of the stereospecific (+)-10-camphorsulfonic acid (10-CSA)-catalyzed cyclization of (*R*)- and (*S*)-enantiomers of 2-methyl-5-(2-methyloxiran-2-yl)pentan-2-ol (Compound **9**), which were in turn synthesized from (*R*)- and (*S*)-enantioforms of linalool (Compound **10**), respectively. The latter cyclization reaction proceeds with very high stereocontrol as previously described by Vidari [30,31], who studied this chemical transformation using very similar epoxides as substrates. In addition, both linalool enantiomers are accessible from the chiral pool with different optical purity [32].

The second approach is based on the lipase-mediated resolution of the racemic alcohol **8**, in turn obtained by reduction of the easily available cinenic acid (Compound **11**) [33].

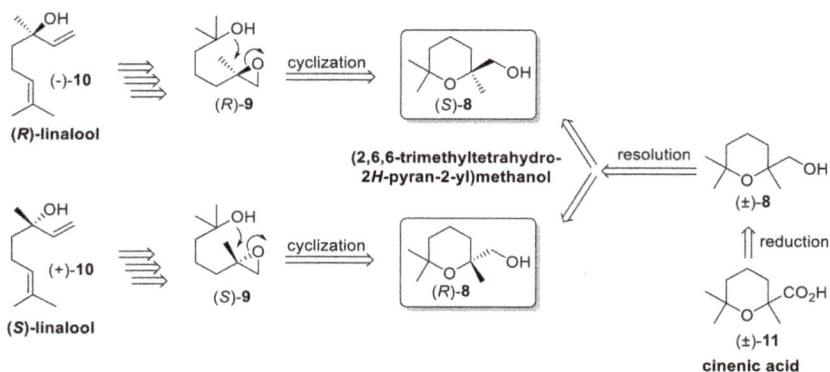

Figure 3. The retrosynthetic analysis of the enantiomeric forms of (2,6,6-trimethyltetrahydro-2*H*-pyran-2-yl)methanol **8**. Two different approaches are proposed: the first is based on stereoselective synthesis (**left**) and the second is based on the resolution of racemic **8** (**right**).

Although we already reported the resolution of racemic **11** by means of the fractional crystallization of its (*R*)-1-phenylethylamine salt [34], the application of this procedure for the preparation of both enantiomers of the aforementioned acid turned out to be lengthy because of the necessary sequential preparation of both (*R*)- and (*S*)-phenylethylamine salts. In order to avoid the tedious chemical manipulations related to the salt formation, as well as to the number of crystallizations required, we investigated the lipase-mediated acetylation reaction of alcohol **8** using vinyl acetate as an acetyl donor. According to this approach, the enzymes are able to catalyze the esterification reaction with high enantioselectivity, allowing the separation of (2,6,6-trimethyltetrahydro-2*H*-pyran-2-yl)methanol isomers. One enantiomer gives the corresponding acetate, and the other does not react. A simple chromatographic separation is required to complete the resolution procedure.

In the present work, we describe the accomplishment of the two above-described synthetic approaches. More specifically, we report in detail the synthetic procedure that allows transforming enantioenriched linalool into enantiopure (2,6,6-trimethyltetrahydro-2*H*-pyran-2-yl)methanol enantiomers. Moreover, we describe a large-scale resolution procedure that exploits the opposite enantioselectivity of two different lipases in the acetylation reaction of alcohol **8**. Thanks to the sequential use of both enzymes, the two enantiomeric forms of the alcohol were obtained in very high enantiomeric purity and were employed for the first stereoselective synthesis of both enantiomers of the natural flavor, linaloyl oxide (2,2,6-trimethyl-6-vinyltetrahydro-2*H*-pyran).

2. Results and Discussion

As mentioned in the introduction, we selected epoxide **9** as a synthetic precursor of alcohol **8**. We planned the enantioselective synthesis of both epoxide enantiomers using linalool enantiomers as starting materials. From a synthetic standpoint, the different reactivity of the two double bonds present in the linalool framework allows an outline of the regioselective preparation of **9**. Indeed, the tertiary hydroxy group can be introduced via regioselective epoxidation/reduction of the trisubstituted linalool double bond (Figure 4). Furthermore, the epoxide functional group can be introduced by oxidative cleavage of the monosubstituted linalool double bond followed by reduction to the corresponding 1,2-diol. The latter intermediate can be transformed into epoxide **9** by activation of the primary hydroxy group followed by a ring-closure reaction.

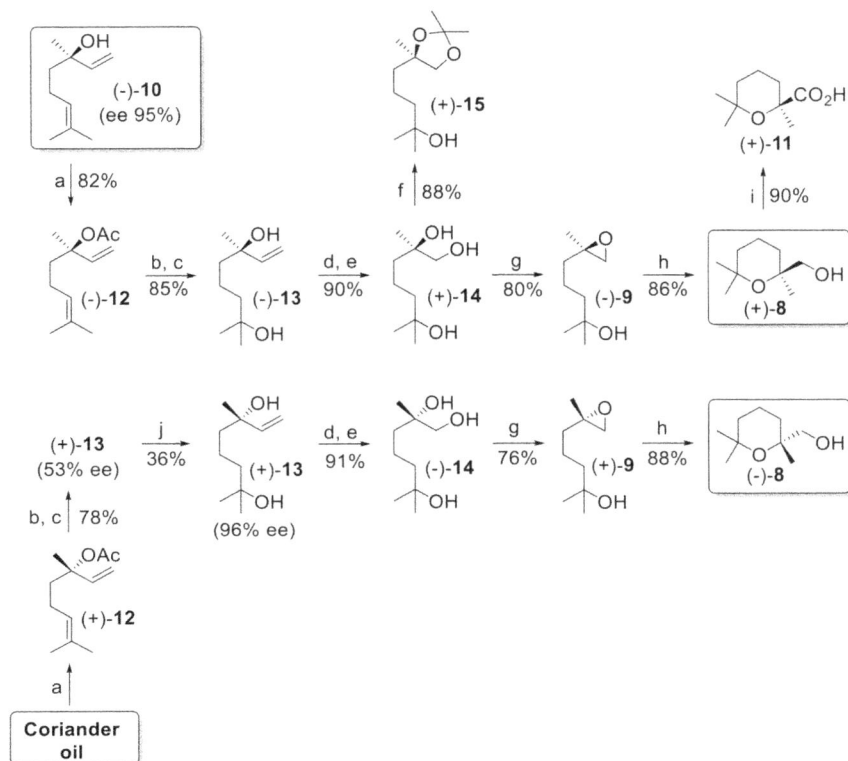

Figure 4. Stereoselective synthesis of (*S*)- and (*R*)-2,6,6-trimethyltetrahydro-2*H*-pyran-2-yl)methanol **8** starting from (*R*)- and (*S*)-linalool **10**, respectively. Reagents and conditions: (**a**) Ac$_2$O/pyridine (Py), 4-dimethylaminopyridine (DMAP) catalyst, reflux 2 h; (**b**) *m*-chloroperbenzoic acid (*m*CPBA), CH$_2$Cl$_2$, 0 °C; (**c**) LiAlH$_4$, dry tetrahydrofuran (THF), reflux 5 h; (**d**) O$_3$, DMAP, −15 °C, then TsCl 0 °C, 3 h; (**h**) (+)-10-camphorsulfonic acid (CSA) catalyst, CH$_2$Cl$_2$, 0 °C, 2 h; (**i**) CrO$_3$/aqueous H$_2$SO$_4$, acetone, 0 °C to room temperature (rt), 1 h; (**j**) three crystallizations from hexane.

At first, we accomplished this synthetic approach starting from a sample of (−)-(*R*)-linalool that is commercially available in very high enantiomeric purity by extraction from *Cinnamomun canphora* (95% enantiomeric excess (ee)). Accordingly, the tertiary alcohol (−)-**10** was refluxed with acetic anhydride and pyridine (Py) in presence of catalytic amount of 4-dimethylaminopyridine (DMAP), to give the corresponding acetate (−)-**12**. The latter ester was treated with *m*-chloroperbenzoic acid (*m*CPBA) in CH$_2$Cl$_2$ and the resulting crude epoxide was regioselectively reduced using an excess of LiAlH$_4$ in refluxing tetrahydrofuran (THF). The obtained diol (−)-**13** was first purified by crystallization from hexane and then treated with ozone in CH$_2$Cl$_2$/MeOH, until complete cleavage of the double bond. The following reduction with NaBH$_4$ afforded the triol (+)-**14** in very good yield. Compound **14** could not be characterized by NMR analysis as the strong inter- and intramolecular hydrogen bonding broadened both ^{1}H- and ^{13}C-NMR signals to such an extent that we could not properly describe the spectra. Since the latter compound is previously undescribed, we prepared the dioxolane (+)-**15** by reaction of (+)-**14** with 2,2-dimethoxypropane (2,2-DMP) in acetone in presence of a pyridinium *p*-toluenesulfonate (PPTS) catalyst. Thus, the derivative (+)-**15** was fully characterized, and the analytical data confirmed both the regioisomeric purity and the chemical structure of triol **14**. The transformation of the 1,2-diol functional group into an epoxide group was performed as described by Vidari [30], using lithium diisopropylamide (LDA) and tosyl chloride in THF solution at −10 °C. The obtained epoxy-alcohol (−)-**9** was treated with a catalytic amount of (+)-10-CSA to give the desired (2,6,6-trimethyltetrahydro-2*H*-pyran-2-yl)methanol (+)-**8** in good yield. The absolute configuration of (+)-**8** was confirmed to be (*S*) by chemical correlation with cinenic acid. Accordingly, oxidation of (+)-**8** with Jones reagent afforded (+)-cinenic acid **11** of known (*S*) configuration [35].

In order to also obtain the compound (−)-**8**, we accomplished the identical reaction sequence described above starting from (+)-linalool. The latter compound is easily available from coriander oil whose weight is made up of more than 90% (+)-**10**. The direct treatment of the essential oil with acetic anhydride and pyridine in presence of a catalytic amount of DMAP, followed by purification by distillation, gave the corresponding acetate (+)-**12** in good yield. Unfortunately, the enantiomeric purity of (+)-linalool from this botanic source ranges from 45 to 85% ee [32], which is not suitable for the preparation of the chiral building block **8**. Since our synthesis makes use of the intermediate 2,6-dimethyloct-7-ene-2,6-diol **13**, which can be purified by crystallization, we observed that, starting from (−)-(*R*)-linalool showing 95% ee, the optical rotation value of the resulting diol was −9.9° which increased to −10.4° after crystallization. Similarly, diol (+)-**13** derived from a commercial sample of coriander oil (optical purity of the linalool of about 55% ee) showed an optical rotation value of +5.2°, which increased to +10.1° after three crystallizations from hexane. These results demonstrate that the enantiopurity of diol **13** was improved through fractional crystallization passing from 53% ee to 96% ee. Therefore, the described purification procedure allows the proper enantioselective synthesis of both enantioforms of compound **8**.

Accordingly, the achieved enantiopure diol (+)-**13**, was transformed into alcohol (−)-**8**, following the experimental conditions used for the synthesis of alcohol (+)-**8**. Overall, both enantioforms of the building block (2,6,6-trimethyltetrahydro-2*H*-pyran-2-yl)methanol were achieved in optical purity superior to 95% ee.

The second part of this work, namely the development of an enzyme-mediated resolution procedure of alcohol (±)-**8**, first requires a valuable amount of the aforementioned racemic alcohol. Hence, we devised the preparative procedure described in Figure 5, which improved the cinenic acid synthesis previously reported by Rupe and Lang [33].

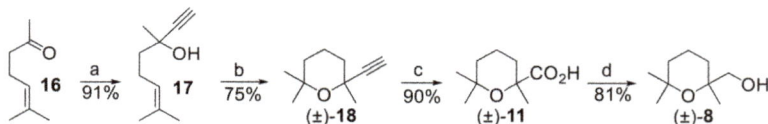

Figure 5. Synthesis of racemic 2,6,6-trimethyltetrahydro-2*H*-pyran-2-yl)methanol **8** starting from 6-methylhept-5-en-2-one **16**. Reagents and conditions: (**a**) ethynylmagnesium bromide, THF dry, −70 °C to rt, 2 h; (**b**) HCOOH/H_2O 85:15, reflux 30 min; (**c**) $NaIO_4$, $RuCl_3 \cdot H_2O$ catalyst, CCl_4/CH_3CN/H_2O 2:2:3, rt, 24 h; (**d**) ClCOOEt/Et_3N, dry THF, −10 °C, one hour, then $NaBH_4$, H_2O, 0 °C to rt, 4 h.

Accordingly, the addition of ethynylmagnesium bromide to 6-methylhept-5-en-2-one (Compound **16**) afforded dehydrolinalool (Compound **17**), which was then heated at reflux in formic acid/water to give 2-ethynyl-2,6,6-trimethyltetrahydro-2*H*-pyran (Compound **18**).

The oxidation of the alkyne **18** to cinenic acid **11**, performed using Rupe and Lang procedure, makes use of aqueous $KMnO_4$, and the yield of the isolated acid does not exceed 50%, even with complete conversion of the alkyne. Therefore, we employed a different oxidation method based on the use of $NaIO_4$, in the presence of a catalytic amount of $RuCl_3 \cdot H_2O$ [36]. As a result, cinenic acid was obtained in very good yield (90%). Finally, the transformation of the carboxylic acid functional group of **11** into the corresponding carboxy-ethyl derivative followed by reduction with $NaBH_4$ [37] smoothly afforded racemic **8**, in good overall yield.

To the best of our knowledge, no enzyme-mediated resolution procedures of (2,6,6-trimethyltetrahydro-2*H*-pyran-2-yl)methanol are reported until now. It should be noted that, usually, lipases catalyze the esterification of primary alcohols with very low enantioselectivities. Since we already described some remarkable exceptions to this behavior [38–41], we decided to investigate the reactivity of racemic **8** in the latter reaction, using a number of selected lipases. For each experiment, **8** was treated with vinyl acetate in *t*-butyl-methyl ether, in the presence of the given enzyme (Figure 6). The reactions were interrupted when the wanted conversion was achieved. The unreacted alcohol and the acetate **19** were separated by chromatography, and their enantiomeric compositions were determined by comparing their optical rotation values with those measured for the same enantiopure chemical compounds. The results are summarized in Table 1.

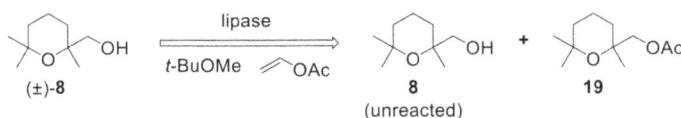

Figure 6. Lipase-mediated acetylation of racemic (2,6,6-trimethyltetrahydro-2*H*-pyran-2-yl)methanol **8**. Reagents and conditions: *t*-BuOMe/vinyl acetate 5:1, rt, concentration of (±)-**8**: 1 M. Results are collected in Table 1.

Table 1. Results of the enzyme-mediated acetylation of racemic alcohol **8**.

Racemic Alcohol	Enzyme	Acetate/Configuration (Enantiomeric Excess (ee))	Time (Days)	Conversion (%)	Enantiomer Ratio (E) [2]
(±)-**8**	PS (Amano)	(−)-**19**/2*R* (61% ee)	3	44	6.5
	PPL	-	14	<10	-
	CRL	(+)-**19**/2*S* (4% ee)	0.5 h	29	1.1
	Novozym 435	(+)-**19**/2*S* (81% ee)	3	40	16.3
	PLE on Eupergit	[1]	-	-	-
	Lipase AK (Amano)	(−)-**19**/2*R* (79% ee)	4	40	13.7

[1] After 10 days, the thin-layer chromatography (TLC) analysis of the reaction mixture did not detect the presence of the acetate; [2] E = ln(1 − c × (1 + ee$_p$))/ln(1 − c × (1 − ee$_p$)), [42]. PPL—porcine pancreatic lipase; CRL—lipase from *Candida rugosa*; PLE—pig liver esterase.

The perusal of the obtained data allows drawing some relevant considerations. Firstly, we can observe that porcine pancreatic lipase (PPL) and pig liver esterase (PLE on Eupergit) were completely inactive. The reaction with PPL afforded only a trace of the acetate **19** after a long reaction time. This behavior is in sharp contrast with the results obtained in our previous studies where the same enzymes showed a remarkable catalytic activity in the acetylation of different primary alcohols. On the contrary, all the other enzymes evaluated in this work showed high activity, although with different stereoselectivity. More specifically, lipase PS and lipase AK (both from Amano Pharmaceuticals) catalyze the acetylation of (*R*)-**8** in modest (E = 6.5) and good (E = 13.7) enantioselectivities, respectively. Moreover, lipase from *Candida rugosa* (CRL) and Novozym 435 lipase catalyze the acetylation of (*S*)-**8** in very low (E = 1.1) and good (E = 16.3) enantioselectivities, respectively. It is worth noting that CRL is a very effective catalyst, but does not show any enantioselectivity, thus making the enzyme unsuitable for any resolution process. Overall, we identified two enzymes, namely Novozym 435 and lipase AK, that are able to catalyze the acetylation of racemic alcohol **8** with good enantioselectivity. Both enzymes are suitable for setting up a resolution process, but they display opposite enantiopreference. Taking advantage of this observation, we devised a large-scale resolution procedure (Figure 7) based on the combined and sequential use of Novozym 435 lipase and lipase AK.

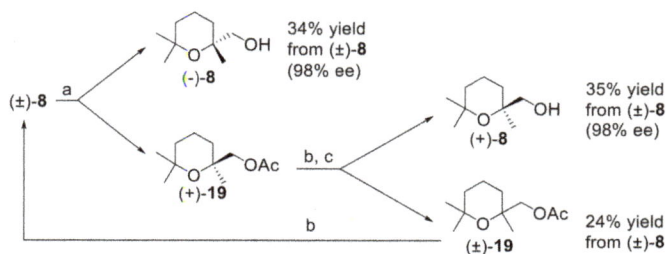

Figure 7. Large-scale resolution procedure of racemic 2,6,6-trimethyltetrahydro-2*H*-pyran-2-yl)methanol **8**. Reagents and conditions: (**a**) Novozym® 435 lipase, *t*-BuOMe/vinyl acetate 5:1, rt; (**b**) NaOH/MeOH, reflux, 1 h; (**c**) lipase AK, *t*-BuOMe/vinyl acetate 5:1, rt.

Hence, racemic **8** was treated with vinyl acetate in *t*-BuOMe using Novozym 435 as catalyst. The acetylation reaction was prolonged until about 65% conversion was reached. In accord with the specific lipase enantioselectivity and with the general rules of the enzyme-based kinetic resolution of racemic mixtures [42], the unreacted alcohol (−)-(*R*)-**8** was isolated in about 34% yield and in very high enantiopurity (98% ee). On the contrary, the acetate (+)-**19** possessed low optical purity and was hydrolyzed using NaOH in methanol. The resulting alcohol (+)-**8** was submitted to a second acetylation step, using lipase AK as catalyst. Indeed, the latter enzyme catalyzes the esterification of the (−)-(*R*) enantiomer of alcohol **8**, which was transformed in the corresponding acetate. As (−)-**8** is the minor component of the enantiomers mixture, the enzymatic reaction increased the enantiomeric purity of the unreacted alcohol. After the acetylation reaction reached a conversion of about 60%, the alcohol (+)-**8** was isolated in about 35% overall yield and with 98% ee. Nearly racemic acetate **19** was also obtained and it could be hydrolyzed to recover further alcohol, to be used in a new resolution procedure.

Overall, our resolution procedure proved to be compact, effective, and user-friendly as it can afford both enantiomeric forms of (2,6,6-trimethyltetrahydro-2*H*-pyran-2-yl)methanol in very high optical purity by means of two commercial enzymes and without the employment of demanding experimental conditions or reagents.

As a first synthetic application of the obtained chiral building blocks, we describe here the preparation of (*R*)- and (*S*)-linaloyl oxide **4** starting from (*R*)- and (*S*)-(2,6,6-trimethyltetrahydro-2*H*-pyran-2-yl)methanol **8**, respectively. As mentioned in the introduction, this monoterpene is a relevant fragrance/flavor ingredient [43]. More specifically, it is one of the fragrance components of geranium and lime essential oil [7,8] and is a trace component of the flavor of many fruits or other vegetal species [9–20]. Since an effective and cheap synthesis of racemic **4** was established [44], linaloyl oxide is currently produced and commercialized in this form, under the trade name Limetol® (Givaudan).

We are not aware of any olfactory evaluation of the single enantiomeric forms of this monoterpene that seems to occur in essential oils in nearly racemic form [8]. Since the organoleptic evaluation of this compound is subject to the availability of both (*R*) and (*S*) isomers in high enantiomeric purity, their stereoselective synthesis is highly desired. To date, only a lengthy and low-yielding synthesis of (+)-**4** was reported in the course of a study finalized to the characterization of linalool oxide [45]. As the starting compound was (−)-linalool, a similar approach for the preparation of enantiopure (−)-**4** is not applicable, as (+)-linalool is available in low enantiomeric purity. Therefore, we synthesized (*R*)- and (*S*)-linaloyl oxide starting from enantiopure (*R*)- and (*S*)-(2,6,6-trimethyltetrahydro-2*H*-pyran-2-yl)methanol **8**, respectively, as described in Figure 8.

Figure 8. Synthesis of (*R*)- and (*S*)-linaloyl oxide **4** starting from (*R*)- and (*S*)-(2,6,6-trimethyltetrahydro-2*H*-pyran-2-yl)methanol **8**, respectively. Reagents and conditions: (**a**) Py·SO$_3$, dimethyl sulfoxide (DMSO), Et$_3$N, rt, 2 h; (**b**) Zn, CH$_2$I$_2$, Me$_3$Al, dry THF, 0 °C, one hour.

Accordingly, the alcohols (−) and (+)-**8** were oxidized in good yields to the corresponding aldehydes (−) and (+)-**20**, respectively, using Py·SO$_3$ in dimethyl sulfoxide (DMSO)/Et$_3$N [46]. Then, the C$_9$ aldehydes **20** were homologated to the C$_{10}$ ethers **4** through a methylenation reaction. The most used method to perform this transformation, namely the Wittig reaction using triphenylphosphoniummethylene ylide, was not successful, as only the degradation of the starting aldehyde was observed. On the contrary, we found that the reagent obtained by reaction of Zn, CH$_2$I$_2$, and Me$_3$Al [47] effectively converted (−) and (+)-**20** into ether (+) and (−)-**4**, respectively. Overall, the described reaction sequence is high yielding and did not involve loss of optical purity. Since the starting chiral building blocks are enantiopure, the obtained linaloyl oxide enantiomers are suitable for olfactory evaluation, the study of which will be reported in due course.

3. Materials and Methods

3.1. Materials and General Methods

All moisture- and air-sensitive reactions were carried out using dry solvents under a static atmosphere of nitrogen.

All solvents and reagents were of commercial quality and were purchased from Sigma-Aldrich (St. Louis, MO, USA) with the exception of (*S*)-linalool acetate (+)-**12** and dehydrolinalool **17** (3,7-dimethyloct-6-en-1-yn-3-ol), which were prepared by acetylation of coriander oil and by addition of ethynylmagnesium bromide to 6-methylhept-5-en-2-one **16**, respectively.

(−)-(*R*)-Linalool, extracted from *Cinnamomun canphora* (L.) and possessing 99% purity by GC and $[\alpha]_D^{20} = -20.9$ (neat), was purchased from Sigma-Aldrich (lot MKBR2739V).

Coriander oil, $[\alpha]_D^{20} = +8.3$ (neat), was purchased from Sigma-Aldrich (lot MKCC6867) and was used as source of (+)-(*S*)-linalool.

Lipase from *Porcine pancreas* (PPL) type II, Sigma-Aldrich, 147 units/mg; lipase from *Pseudomonas cepacia* (PS), Amano Pharmaceuticals Co., Tokyo, Japan, 30 units/mg; lipase from *Candida rugosa* (CRL) type VII, Sigma-Aldrich, ≥700 units/g; Novozym® 435, Novozymes, ≥5000 units/g; lipase AK Amano from *Pseudomonas fluorescens*, Sigma-Aldrich, 20 units/mg; and PLE on Eupergit, Sigma-Aldrich, 200 units/g were employed in this work.

3.2. Analytical Methods and Characterization of the Chemical Compounds

^{1}H and ^{13}C-NMR spectra and DEPT (Distortionless enhancement by polarization transfer) experiments: CDCl$_3$ solutions at room temperature (rt) using a Bruker-AC-400 spectrometer (Billerica, MA, USA) at 400, 100, and 100 MHz, respectively; ^{13}C spectra are proton-decoupled; chemical shifts in ppm relative to internal SiMe$_4$ (0 ppm).

Thin-layer chromatography (TLC) involved the use of Merck silica gel 60 F$_{254}$ plates (Merck Millipore, Milan, Italy), while column chromatography involved the use of silica gel.

Melting points were measured on a Reichert apparatus, equipped with a Reichert microscope, and are uncorrected.

Optical rotations were measured on a Jasco-DIP-181 digital polarimeter (Jasco, Tokyo, Japan).

Mass spectra were recorded on a Bruker ESQUIRE 3000 PLUS spectrometer (ESI detector, Billerica, MA, USA) or by GC–MS analyses.

GC–MS analyses involved the use of an HP-6890 gas chromatograph equipped with a 5973 mass detector, using an HP-5MS column (30 m × 0.25 mm, 0.25-μm film thickness; Hewlett Packard, Palo Alto, CA, USA) with the following temperature program: 60° (1 min), then 6°/min to 150° (held at 1 min), then 12°/min to 280° (held 5 min); carrier gas: He; constant flow 1 mL/min; split ratio: 1/30; t_R given in minutes.

The values of t_R for each compound are as follows: t_R(**4**) 5.97, t_R(**8**) 8.78, t_R(**9**) 10.53, t_R(**10**) 8.86, t_R(**11**) 12.32, t_R(**12**) 12.42, t_R(**13**) 11.93, t_R(**15**) 14.58, t_R(**17**) 8.67, t_R(**18**) 5.81, t_R(**19**) 12.63, and t_R(**20**) 7.07.

3.3. Stereoselective Preparation of (S)- and (R)-(2,6,6-Trimethyltetrahydro-2H-pyran-2-yl)methanol Starting from (R)- and (S)-Linalool, Respectively

3.3.1. (R)-2,6-Dimethyloct-7-ene-2,6-diol (−)-**13**

(−)-(R)-Linalool **10** (60 g, 389 mmol), acetic anhydride (50 mL, 529 mmol), pyridine (45 mL, 559 mmol), and DMAP (1 g, 8.2 mmol) were heated at reflux under a static atmosphere of nitrogen until complete acetylation of the starting alcohol (2 h by TLC analysis). The reaction mixture was then cooled and quenched by addition to a mixture of water and crushed ice followed by extraction with diethyl ether (2 × 250 mL). The combined organic phases were washed in turn with water, and saturated with NaHCO$_3$ solution (3 × 200 mL) and brine. The organic phase was dried (Na$_2$SO$_4$) and concentrated in vacuo. The residue was purified by distillation to afford colorless (R)-linalool acetate **12** (62.9 g, 82% yield, 94% purity by GC analysis).

A sample of acetate **12** (24 g, 122 mmol) was dissolved in CH$_2$Cl$_2$ (100 mL) and treated with *m*-chloroperbenzoic acid (31 g, 77% *w/w*, 138 mmol) stirring at 0 °C until completion of the reaction (TLC monitoring). The *m*-chlorobenzoic acid formed was removed by filtration and the liquid phase was washed in turn with aqueous Na$_2$SO$_3$ (10% *w/w*), aqueous NaHCO$_3$ (saturated solution) and brine. The organic phase was dried (Na$_2$SO$_4$), concentrated under reduced pressure and the residue was diluted with dry THF (60 mL). The aforementioned solution was added dropwise to a stirred suspension of LiAlH$_4$ (6 g, 158 mmol) in dry THF (200 mL). The reaction was stirred at reflux until the starting epoxide was completely transformed into the diol **13** (5 h, TLC analysis); then, the mixture was cooled (0 °C), diluted with diethyl ether (300 mL), and quenched by dropwise addition of 40% aqueous solution of NaOH (50 mL), stirring vigorously for 1 h. The resulting heterogeneous mixture was filtered on a celite pad and the organic phase was washed with brine, before being dried (Na$_2$SO$_4$) and concentrated under reduced pressure. The residue was purified by chromatography using *n*-hexane/AcOEt (8:2–1:1) as an eluent to afford pure (R)-2,6-dimethyloct-7-ene-2,6-diol **13** (17.8 g, 85% yield) as a colorless oil which solidified on standing; $[\alpha]_D^{20} = -9.9$ (*c* 2, CHCl$_3$). A sample of the diol was recrystallized from hexane. The collected crystals showed 98% purity by GC analysis; melting point (mp): 55–56 °C; $[\alpha]_D^{20} = -10.4$ (*c* 2.1, CHCl$_3$); ^{1}H NMR (400 MHz, CDCl$_3$): δ 5.92 (dd, *J* = 17.4, 10.8 Hz, 1H), 5.21 (dd, *J* = 17.4, 1.2 Hz, 1H), 5.05 (dd, *J* = 10.8, 1.2 Hz, 1H), 1.69 (br s, 2H), 1.59–1.49 (m, 2H), 1.49–1.34 (m, 4H), 1.29 (s, 3H), 1.21 (s, 6H). ^{13}C NMR (100 MHz, CDCl$_3$): δ 145.1 (CH), 111.6 (CH$_2$), 73.2 (C), 71.0 (C), 44.1 (CH$_2$), 42.6 (CH$_2$), 29.2 (Me), 27.7 (Me), 18.6 (CH$_2$).

GC–MS *m/z* (relative intensity): 154 (M$^+$-H$_2$O, <1), 139 (11), 121 (19), 109 (8), 93 (10), 81 (37), 71 (100), 59 (33), 43 (46).

3.3.2. (R)-2,6-Dimethylheptane-1,2,6-triol (+)-**14**

An oxygen stream containing ozone (0.1 mole/hour) was bubbled into a cooled (−70 °C) solution of the diol (−)-**13** (14 g, 81.3 mmol) in CH$_2$Cl$_2$/MeOH (3:1 *v/v*, 280 mL). As soon as the solution took a persistent light-blue color, the ozone addition was stopped and the oxygen stream was switched to a nitrogen stream. After a few minutes the solution became pale yellow and NaBH$_4$ (5 g, 132.2 mmol) was added portionwise. The reaction was slowly allowed to reach rt and set aside overnight. The excess of hydride was then quenched by addition of acetic acid (50 mL). After excluding the presence of residual peroxides (negative KI/starch test) the solvents and the excess of acetic acid were removed under reduced pressure and the residue was diluted with water (80 mL) and extracted with *n*-butanol (4 × 100 mL). The combined organic phases were washed with brine, before being dried (Na_2SO_4) and concentrated under reduced pressure. The residue was purified by chromatography eluting first with *n*-hexane/AcOEt (1:1) and then increasing the polarity using AcOEt/MeOH (2:1) to afford pure (+)-(*R*)-2,6-dimethylheptane-1,2,6-triol **14** as a thick, colorless oil (12.9 g, 90% yield).

$[\alpha]_D^{20}$ = +2.9 (*c* 4.7, CHCl$_3$); MS (ESI): 199.1 (M$^+$ + Na)

Due to the strong intermolecular hydrogen bonding, ^{1}H and ^{13}C-NMR analysis of the latter compound were not clear and did not allow an unambiguous characterization of compound **14**. Therefore, we prepared the derivative **15** that was fully characterized. The obtained analytical data substantiated the chemical structure of **15**, and thus, the structure of triol **14**. Accordingly, a sample of triol (+)-**14** (200 mg, 1.13 mmol) was dissolved in acetone (10 mL) and treated with 2,2-dimethoxypropane (5 mL) and pyridinium *p*-toluenesulfonate (0.1 g, 0.40 mmol), and then stirred at rt for 8 h. Next, Et$_3$N (2 mL) was added and the solvents were removed under reduced pressure. The residue was partitioned between EtOAc (50 mL) and water (50 mL) and the organic phase was washed with water and with brine, before being dried (Na_2SO_4) and concentrated in vacuo. The obtained pale-yellow oil was purified by chromatography using *n*-hexane/AcOEt (8:2–1:1) as an eluent to afford pure (+)-(*R*)-2-methyl-5-(2,2,4-trimethyl-1,3-dioxolan-4-yl)pentan-2-ol **15** (215 mg, 88%).

$[\alpha]_D^{20}$ = +2.3 (*c* 3.2, CHCl$_3$)

^{1}H NMR (400 MHz, CDCl$_3$): δ 3.79 (d, *J* = 8.3 Hz, 1H), 3.71 (d, *J* = 8.3 Hz, 1H), 1.72–1.33 (m, 7H), 1.40 (s, 3H), 1.38 (s, 3H), 1.28 (s, 3H), 1.22 (s, 6H).

^{13}C NMR (100 MHz, CDCl$_3$): δ 109.0 (C), 81.2 (C), 74.0 (CH$_2$), 70.9 (C), 44.3 (CH$_2$), 40.5 (CH$_2$), 29.3 (Me), 29.2 (Me), 27.2 (Me), 27.1 (Me), 24.8 (Me), 19.3 (CH$_2$).

GC–MS *m*/*z* (relative intensity): 201 (M$^+$-Me, 31), 183 (16), 141 (9), 123 (100), 115 (92), 107 (10), 97 (19), 81 (39), 72 (52), 59 (50), 43 (75).

MS (ESI): 239.1 (M$^+$ + Na).

3.3.3. (*R*)-2-Methyl-5-(2-methyloxiran-2-yl)pentan-2-ol (−)-**9**

A stirred solution of triol (+)-**14** (8.2 g, 46.5 mmol) in dry THF (60 mL) was treated at −15 °C with freshly prepared LDA (44 mL of a 2.4 M solution in THF). After ten minutes, a solution of tosyl chloride (9.7 g, 50.9 mmol) in dry THF (30 mL) was added dropwise. The mixture was stirred at 0 °C until complete transformation of the starting triol (3 h). Then, the reaction was quenched by pouring into a mixture of saturated NH$_4$Cl solution and crushed ice followed by extraction with diethyl ether (2 × 200 mL). The combined organic phases were washed in turn with saturated NaHCO$_3$ solution and brine. The organic solution was dried (Na_2SO_4) and concentrated in vacuo. The residue was purified by chromatography using *n*-hexane/AcOEt (9:1–1:1) as an eluent to afford pure (−)-(*R*)-2-methyl-5-(2-methyloxiran-2-yl)pentan-2-ol **9** as a colorless oil (5.9 g, 80% yield).

$[\alpha]_D^{20}$ = −6.2 (*c* 5.2, CHCl$_3$)

^{1}H NMR (400 MHz, CDCl$_3$): δ 2.62 (d, *J* = 4.8 Hz, 1H), 2.58 (d, *J* = 4.8 Hz, 1H), 1.67–1.41 (m, 7H), 1.32 (s, 3H), 1.22 (s, 6H).

^{13}C NMR (100 MHz, CDCl$_3$): δ 70.7 (C), 56.9 (C), 53.8 (CH$_2$), 43.6 (CH$_2$), 36.9 (CH$_2$), 29.3 (Me), 29.1 (Me), 20.8 (Me), 19.9 (CH$_2$).

MS (ESI): 181.1 (M$^+$ + Na).

3.3.4. (*S*)-(2,6,6-Trimethyltetrahydro-2*H*-pyran-2-yl)methanol (+)-**8**

A stirred solution of epoxide (−)-**9** (5.6 g, 35.4 mmol) in CH_2Cl_2 (50 mL) was treated at 0 °C with (+)-10-camphorsulfonic acid (100 mg, 0.43 mmol). As soon as the starting epoxide was no longer detectable by TLC analysis (2 h), the reaction was quenched by addition of a saturated $NaHCO_3$ solution (40 mL) and was extracted with CH_2Cl_2 (2 × 60 mL). The combined organic phases were dried (Na_2SO_4) and concentrated in vacuo. The residue was purified by chromatography using *n*-hexane/Et$_2$Ot (9:1–2:1) as an eluent to afford pure (+)-(*S*)-(2,6,6-trimethyltetrahydro-2*H*-pyran-2-yl)methanol **8** as a colorless oil (4.8 g, 86% yield).

$[\alpha]_D^{20}$ = +9.8 (*c* 3.4, CHCl$_3$)

^{1}H NMR (400 MHz, CDCl$_3$): δ 3 32 (d, *J* = 10.6 Hz, 1H), 3.24 (d, *J* = 10.6 Hz, 1H), 2.28 (br s, 1H), 1.84–1.58 (m, 3H), 1.55–1.47 (m, 1H), 1.41–1.25 (m, 2H), 1.25 (s, 3H), 1.19 (s, 3H), 1.17 (s, 3H).

^{13}C NMR (100 MHz, CDCl$_3$): δ 73.5 (C), 71.8 (C), 70.6 (CH$_2$), 36.4 (CH$_2$), 32.2 (Me), 30.3 (CH$_2$), 28.1 (Me), 24.3 (Me), 16.1 (CH$_2$).

GC–MS *m/z* (relative intensity): 143 (M$^+$-Me, 2), 127 (59), 109 (100), 97 (3), 81 (4), 75 (4), 69 (49), 59 (12), 43 (45).

3.3.5. Chemical Correlation of (+)-(2,6,6-Trimethyltetrahydro-2*H*-pyran-2-yl)methanol **8** with (+)-(*S*)-Cinenic Acid **11**

Jones reagent (5 mmol) was added dropwise to a stirred solution of the alcohol (+)-**8** (0.2 g, 1.26 mmol, 95% ee) in acetone (15 mL) at 0 °C. The reaction was allowed to reach rt and stirring was prolonged since TLC analysis indicated complete transformation of the intermediate aldehyde into the corresponding acid (one hour). The reaction was then quenched by dilution with water (60 mL) and extraction with diethyl ether (2 × 70 mL). The organic phase was washed with water and with brine, before being dried (Na_2SO_4), and concentrated under reduced pressure. The residue was purified by chromatography using *n*-hexane/ethyl acetate (9:1–7:3) as an eluent to give pure (+)-(*S*)-2,6,6-trimethyltetrahydro-2*H*-pyran-2-carboxylic acid **11** as a colorless oil which crystallized on standing (195 mg, 90% yield, 94% purity by GC–MS analysis).

$[\alpha]_D^{20}$ = +2.6 (*c* 4, CHCl$_3$)

^{1}H NMR (400 MHz, CDCl$_3$): δ 9.95 (br s, 1H), 2.11–2.02 (m, 1H), 1.80–1.62 (m, 2H), 1.56–1.47 (m, 3H), 1.45 (s, 3H), 1.28 (s, 3H), 1.24 (s, 3H).

^{13}C NMR (100 MHz, CDCl$_3$): δ 178.5 (C), 75.2 (C), 74.4 (C), 35.9 (CH$_2$), 32.0 (CH$_2$), 30.1 (Me), 27.7 (Me), 27.4 (Me), 16.4 (CH$_2$).

GC–MS *m/z* (relative intensity): 157 (M$^+$-Me, 2), 139 (4), 127 (58), 109 (100), 95 (3), 69 (62), 59 (14), 43 (57).

3.3.6. (*S*)-2,6-Dimethyloct-7-ene-2,6-diol (+)-**13**

Coriander oil (60 g), acetic anhydride (50 mL, 529 mmol), pyridine (45 mL, 556 mmol), and DMAP (1 g, 8,2 mmol) were heated at reflux under a static atmosphere of nitrogen until complete acetylation of the (+)-(*S*)-linalool (2 h by TLC analysis). The reaction mixture was then cooled and quenched by addition to a mixture of water and crushed ice followed by extraction with diethyl ether (2 × 200 mL). The combined organic phases were washed in turn with water, before being saturated $NaHCO_3$ solution (3 × 200 mL) and brine. The organic phase was dried (Na_2SO_4) and concentrated in vacuo. The residue was purified by distillation to afford colorless (+)-(*S*)-linalool acetate **12** (51.2 g, 91% purity by GC analysis).

A sample of acetate (+)-**12** (28 g, 142.6 mmol) was dissolved in CH_2Cl_2 (120 mL) and treated with *m*-chloroperbenzoic acid (34 g, 77% *w/w*, 151.7 mmol) stirring at 0 °C until completion of the reaction (TLC monitoring). The *m*-chlorobenzoic acid formed was removed by filtration and the liquid phase was washed in turn with aqueous Na_2SO_3 (10% w/w), aqueous $NaHCO_3$ (saturated solution), and brine. The organic phase was dried (Na_2SO_4), concentrated under reduced pressure and the residue was diluted with dry THF (60 mL). The aforementioned solution was added dropwise to a stirred suspension of $LiAlH_4$ (6.5 g, 171.3 mmol) in dry THF (200 mL). The reaction was stirred at reflux until the starting epoxide was completely transformed in the diol (+)-**13** (5 h, TLC analysis); then, the mixture was cooled (0 °C), diluted with diethyl ether (300 mL), and quenched by dropwise addition of 40% aqueous solution of NaOH (50 mL), stirring vigorously for 1 h. The resulting heterogeneous mixture was filtered on a celite pad and the organic phase was washed with brine, before being dried (Na_2SO_4) and concentrated under reduced pressure. The residue was purified by chromatography using *n*-hexane/AcOEt (8:2–1:1) as an eluent to afford pure (+)-(*S*)-2,6-dimethyloct-7-ene-2,6-diol **13** (19.1 g, 78% yield) as a colorless thick oil, showing $[\alpha]_D^{20}$ = +5.3 (*c* 3.5, CHCl$_3$). The diol was then recrystallized three times from hexane. The third crystal crop (6.8 g, recrystallization yield 36%) showed 98% purity by GC analysis; mp: 52–53 °C; $[\alpha]_D^{20}$ = +10.1 (*c* 3.5, CHCl$_3$), corresponding to 96% ee. ^{1}H-NMR, ^{13}C-NMR, and GC–MS were superimposable to those reported for the (−)-(*R*) isomer.

3.3.7. (*S*)-2,6-Dimethylheptane-1,2,6-triol (−)-**14**

According to the procedure outlined for the synthesis of triol (+)-**14**, diol (+)-**13** (96% ee, 98% chemical purity) gave, in 91% yield, (−)-**14** as a colorless thick oil with $[\alpha]_D^{20}$ = −2.8 (*c* 3.5, CHCl$_3$).

3.3.8. (*S*)-2-Methyl-5-(2-methyloxiran-2-yl)pentan-2-ol (+)-**9**

According to the procedure outlined for the synthesis of epoxide (−)-**9**, triol (−)-**14** (96% ee) gave, in 76% yield, (+)-**9** as a colorless oil with $[\alpha]_D^{20}$ = +6.0 (*c* 4.1, CHCl$_3$) and 95% chemical purity by GC; ^{1}H-NMR, ^{13}C-NMR, and MS (ESI) were superimposable to those reported for the (−)-(*R*) isomer.

3.3.9. (*R*)-(2,6,6-Trimethyltetrahydro-2*H*-pyran-2-yl)methanol (−)-**8**

According to the procedure outlined for the synthesis of alcohol (+)-**8**, epoxide (+)-**9** (96% ee, 95% chemical purity) gave, in 88% yield, (−)-**8** as a colorless oil with $[\alpha]_D^{20}$ = −10.0 (*c* 3.1, CHCl$_3$) and 95% chemical purity by GC.

3.4. Synthesis of Racemic (2,6,6-Trimethyltetrahydro-2H-pyran-2-yl)methanol

3.4.1. 2-Ethynyl-2,6,6-trimethyltetrahydro-2*H*-pyran

A solution of ethynylmagnesium bromide (390 mL, 0.9 M in THF) was added dropwise at −70 °C to a stirred solution of 6-methylhept-5-en-2-one **16** (40 g, 317 mmol) in dry THF (100 mL) under a static atmosphere of nitrogen. The reaction was allowed to reach rt, and after 2 h, was poured into a mixture of crushed ice (300 g) and saturated NH_4Cl solution (300 mL) followed by extraction with diethyl ether (2 × 300 mL). The combined organic phases were washed with water and with brine, before being dried (Na_2SO_4) and concentrated in vacuo. The residue was purified by distillation (boiling point (bp) 98 °C at 20 mmHg) to afford pure 3,7-dimethyloct-6-en-1-yn-3-ol **17** (43.8 g, 91% yield, 95% purity by GC–MS analysis).

A solution of the alkynol **17** (20 g, 131.4 mmol) in $HCOOH/H_2O$ (85:15, 25 mL) under a static atmosphere of nitrogen was heated at reflux until complete transformation of the starting propargylic alcohol (half an hour, TLC analysis). After cooling, the reaction was diluted with cool water (250 mL) and extracted with diethyl ether (2 × 150 mL). The combined organic phases were washed in turn with water, with saturated $NaHCO_3$ solution and with brine, before being dried (Na_2SO_4) and concentrated in vacuo. The residue was purified by distillation (bp 65 °C at 20 mmHg) to afford pure 2-ethynyl-2,6,6-trimethyltetrahydro-2*H*-pyran **18** (15.1 g, 75% yield, 94% purity by GC–MS analysis) as a colorless oil.

¹H NMR (400 MHz, CDCl$_3$): δ 2.34 (s, 1H), 2.01 (qt, *J* = 13.5, 3.4 Hz, 1H), 1.84 (dm, *J* = 13.0 Hz, 1H), 1.61 (dt, *J* = 13.5, 3.7 Hz, 1H), 1.55 (dm, *J* = 13.0 Hz, 1H), 1.47 (s, 3H), 1.46 (s, 3H), 1.43–1.28 (m, 2H), 1.18 (s, 3H).

¹³C NMR (100 MHz, CDCl$_3$): δ 89.0 (C), 73.5 (C), 71.3 (CH), 67.1 (C), 38.3 (CH$_2$), 36.4 (CH$_2$), 33.0 (Me), 32.6 (Me), 25.2 (Me), 17.5 (CH$_2$).

GC–MS *m/z* (relative intensity): 137 (M$^+$-Me, 100), 119 (11), 109 (76), 95 (32), 79 (69), 66 (81), 56 (62), 43 (83).

3.4.2. 2,6,6-Trimethyltetrahydro-2*H*-pyran-2-carboxylic Acid or Cinenic Acid **11**

A heterogeneous mixture of the alkyne **18** (10 g, 65.7 mmol), sodium periodate (60 g, 280.5 mmol) CCl$_4$ (50 mL), CH$_3$CN (50 mL), water (75 mL), and a catalytic amount of RuCl$_3$ hydrate (40% *w/w* Ru, 80 mg, 0.32 mmol) was vigorously stirred at rt. When the starting alkyne was no longer detectable by TLC analysis (24 h), the reaction was diluted with water (200 mL), acidified using diluted aqueous HCl and extracted with CH$_2$Cl$_2$ (3 × 120 mL). The combined organic phases were dried (Na$_2$SO$_4$) and were concentrated in vacuo. The residue was purified by chromatography using *n*-hexane/AcOEt (9:1–1:1) as an eluent to afford pure 2,6,6-trimethyltetrahydro-2*H*-pyran-2-carboxylic acid **11** as a colorless oil which crystallized on standing (10 2 g, 90% yield, 94% purity by GC–MS analysis). A sample of the acid was recrystallized from hexane. The collected crystals showed 98% purity by GC–MS analysis; mp: 83–84 °C; ¹H-NMR, ¹³C-NMR, and GC–MS were superimposable to those reported above for the (+)-(*S*)-isomer.

3.4.3. Racemic (2,6,6-Trimethyltetrahydro-2*H*-pyran-2-yl)methanol **8**

Ethyl chloroformate (5.5 mL, 57.5 mmol) was added dropwise at −10 °C to a stirred solution of acid **11** (9 g, 52.3 mmol) and Et$_3$N (8 mL, 57.4 mmol) in dry THF (60 mL). After one hour, the precipitate triethylammonium chloride was filtered and the solid was washed with cold THF (10 mL). The combined liquid phases were added dropwise to a stirred solution of NaBH$_4$ (5 g, 132.2 mmol) in water (50 mL) keeping the temperature below 10 °C by external cooling. After complete addition, the reaction was allowed to reach rt, before being stirred at this temperature for 4 h and quenched by acidification with diluted HCl (3% in water). The obtained mixture was extracted with diethyl ether (3 × 100 mL) and the combined organic phases were washed with saturated NaHCO$_3$ solution and with brine, before being dried (Na$_2$SO$_4$) and concentrated in vacuo. The residue was purified by chromatography using *n*-hexane/AcOEt (9:1–7:3) as an eluent to afford pure (2,6,6-trimethyltetrahydro-2*H*-pyran-2-yl)methanol **8** (6.7 g, 81% yield) as a colorless oil; ¹H-NMR, ¹³C-NMR, and GC–MS were superimposable to those reported above for the (+)-(*S*) isomer.

3.5. Enzyme-Mediated Resolution of (2,6,6-Trimethyltetrahydro-2H-pyran-2-yl)methanol

3.5.1. Determination of the Enantioselectivity in the Lipase-Catalyzed Acetylation of Racemic Alcohol **8**

A solution of the racemic alcohol **8** (0.5 g, 3.16 mmol), lipase/esterase, vinyl acetate (5 mL), and *t*-BuOMe (20 mL) was stirred at rt, and the formation of the acetylated compound was monitored by TLC analysis. The reaction was stopped at the reported conversion (see Table 1) by filtration of the enzyme and evaporation of the solvent at reduced pressure. The residue was then purified by chromatography using hexane-acetate (9:1–7:3) as an eluent. The obtained acetate and the unreacted alcohol were bulb-to-bulb distilled in order to obtain solvent free samples suitable for the accurate measurement of their optical rotation values. The enantiomeric purity of the samples was determined comparing the measured optical rotation values with those of (*S*)-(2,6,6-trimethyltetrahydro-2*H*-pyran-2-yl)methanol **8** and (*S*)-(2,6,6-trimethyltetrahydro-2*H*-pyran-2-yl)methanol acetate (+)-**19** possessing 95% ee, $[\alpha]_D^{20}$ = +9.8 (*c* 3.4, CHCl$_3$) and $[\alpha]_D^{20}$ = +7.9 (*c* 2 9, CHCl$_3$), respectively.

Data analysis for acetate (+)-**19**: ^{1}H NMR (400 MHz, CDCl$_3$): δ 4.02 (d, *J* = 10.9 Hz, 1H), 3.86 (d, *J* = 10.9 Hz, 1H), 2.08 (s, 3H), 1.76–1.58 (m, 2H), 1.56–1.32 (m, 4H), 1.23 (s, 3H), 1.21 (s, 3H), 1.18 (s, 3H).

^{13}C NMR (100 MHz, CDCl$_3$): δ 170.9 (C), 71.8 (C), 71.5 (C), 70.9 (CH$_2$), 36.3 (CH$_2$), 31.7 (CH$_2$), 30.8 (Me), 29.7 (Me), 25.4 (Me), 20.9 (Me), 16.0 (CH$_2$).

GC–MS *m/z* (relative intensity): 185 (M$^+$-Me, 2), 127 (71), 109 (100), 97 (3), 81 (3), 69 (36), 56 (6), 43 (54).

3.5.2. Large-Scale Resolution of (2,6,6-Trimethyltetrahydro-2*H*-pyran-2-yl)methanol 8

A solution of the racemic alcohol **8** (10 g, 63.2 mmol), Novozym®435 lipase (3 g), vinyl acetate (10 mL), and t-BuOMe (50 mL) was stirred at rt, and the formation of the acetylated compound was monitored by TLC analysis. The reaction was stopped at 65% conversion by filtration of the enzyme and evaporation of the solvent at reduced pressure. The residue was then purified by chromatography using hexane-acetate (9:1–7:3) as an eluent. The unreacted alcohol (−)-**8** (3.4 g, 34% yield) showed the following analytical data: 96% chemical purity by GC–MS, $[\alpha]_D^{20}$ = −10.1 (*c* 3.6, CHCl$_3$), corresponding to 98% ee. The obtained acetate (+)-**19** was treated with NaOH (6 g, 0.15 mol) in MeOH (80 mL) at reflux for 1 h. After the work-up procedure, the obtained alcohol was submitted again to the resolution procedure using lipase AK (4 g) as a catalyst, vinyl acetate (10 mL), and *t*-BuOMe (50 mL) allowing the acetylation reaction to reach a conversion of about 60%. The unreacted alcohol (+)-**8** (3.5 g, 35% yield) showed the following analytical data: 97% chemical purity by GC–MS, $[\alpha]_D^{20}$ = +10.1 (*c* 3.0, CHCl$_3$), corresponding to 98% ee. The remaining acetate (3.0 g, 24% yield) could be hydrolyzed to recover further alcohol with low ee that could be used in a new resolution procedure.

3.6. Synthesis of the Enantiomeric Forms of 2,2,6-Trimethyl-6-vinyltetrahydro-2H-pyran

3.6.1. (R)-2,6,6-Trimethyltetrahydro-2*H*-pyran-2-carbaldehyde (−)-20

A solution of Py·SO$_3$ complex (2.7 g, 17 mmol) in dry DMSO (10 mL) was added in one portion to a stirred solution of alcohol (−)-**8** (1 g, 6.32 mmol, 98% ee) and Et$_3$N (10 mL, 72 mmol) in dry DMSO (15 mL). After complete transformation of the starting alcohol (by TLC analysis, 2 h), the reaction was quenched by addition of water (100 mL) followed by extraction with diethyl ether (2 × 80 mL). The combined organic phases were washed in turn with water, diluted HCl solution, and brine. The organic solution was dried (Na$_2$SO$_4$) and concentrated in vacuo. The residue was purified by chromatography using *n*-hexane/Et$_2$O (95:5–8:2) as an eluent to afford pure (−)-(R)-2,6,6-trimethyltetrahydro-2*H*-pyran-2-carbaldehyde **20** (810 mg, 82% yield, 92% purity by GC–MS analysis) as a colorless oil.

$[\alpha]_D^{20}$ = −44.7 (*c* 3.9, CHCl$_3$)

^{1}H NMR (400 MHz, CDCl$_3$): δ 9.59 (d, *J* = 1.7 Hz, 1H), 2.15 (dm, *J* = 13.3 Hz, 1H), 1.61–1.35 (m, 4H), 1.30–1.17 (m, 1 H), 1.27 (s, 3H), 1.12 (s, 3H), 1.11 (s, 3H).

^{13}C NMR (100 MHz, CDCl$_3$): δ 205.5 (C), 78.4 (C), 73.0 (C), 35.7 (CH$_2$), 31.9 (Me), 29.2 (CH$_2$), 25.8 (Me), 24.5 (Me), 16.6 (CH$_2$).

GC–MS *m/z* (relative intensity): 141 (M$^+$-Me, 2), 127 (60), 109 (100), 95 (3), 81 (4), 69 (75), 59 (8), 43 (54).

3.6.2. (R)-2,2,6-Trimethyl-6-vinyltetrahydro-2*H*-pyran (+)-4

A solution of trimethylaluminium in hexane (3.2 mL of a 1 M solution) was added dropwise, under a static atmosphere of nitrogen, to a stirred suspension of activated zinc dust (3.2 g, 48.9 mmol), CH$_2$I$_2$ (4.2 g, 15.7 mmol), and dry THF (20 mL). The temperature of the mixture was kept below 30 °C by external cooling until the exothermic reaction settled down. The stirring was prolonged at rt for a further 10 min; then, aldehyde (−)-**20** (0.8 g, 5.12 mmol, 98% ee) in dry THF (3 mL) was added dropwise at 0 °C and the mixture was stirred for one further hour. The reaction was diluted with diethyl ether (100 mL) and was acidified with diluted HCl (3% in water). The ether was separated and the aqueous phase was extracted with further ether (60 mL). The combined organic phases were washed with water and with brine, before being dried (Na_2SO_4) and concentrated in vacuo. The residue was purified by chromatography using *n*-pentane/Et_2O (99:1–9:1) as an eluent to afford pure (+)-2,2,6-trimethyl-6-vinyltetrahydro-2*H*-pyran **4** (0.61 g, 77% yield, 93% purity by GC–MS analysis) as a colorless oil. The bulb-to-bulb distillation (60–65 °C, 20 mmHg) of the latter compound afforded very pure (+)-**4** (99% purity by GC–MS analysis).

$[\alpha]_D^{20}$ = +8.9 (*c* 3.3, $CHCl_3$)

^{1}H NMR (400 MHz, $CDCl_3$): δ 5.96 (ddd, *J* = 17.8, 11.0, 0.8 Hz, 1H), 4.99 (dd, *J* = 17.8, 0.8 Hz, 1H), 4.94 (dd, *J* = 11.0, 1.0 Hz, 1H), 1.89 (dm, *J* = 13.4 Hz, 1H), 1.78–1.65 (m, 1H), 1.62–1.51 (m, 1H), 1.50–1.32 (m, 3H), 1.20 (s, 3H), 1.19 (s, 3H), 1.18 (s, 3H).

^{13}C NMR (100 MHz, $CDCl_3$): δ 147.2 (CH), 110.2 (CH_2), 73.4 (C), 72.2 (C), 36.6 (CH_2), 33.0 (CH_2), 32.3 (Me), 31.5 (Me), 27.5 (Me), 16.8 (CH_2).

GC–MS *m/z* (relative intensity): 154 (M$^+$, <1), 139 (100), 121 (42), 109 (24), 93 (11), 81 (59), 71 (64), 56 (29), 43 (47).

3.6.3. (*S*)-2,6,6-Trimethyltetrahydro-2*H*-pyran-2-carbaldehyde (+)-**20**

According to the procedure outlined for the synthesis of aldehyde (−)-**20**, alcohol (+)-**8** (98% ee) gave, in 85% yield, aldehyde (+)-**20** as a colorless oil with $[\alpha]_D^{20}$ = +45.4 (*c* 3.1, $CHCl_3$) and 95% chemical purity by GC; ^{1}H-NMR, ^{13}C-NMR, and GC–MS were superimposable to those reported for the (−)-(*R*)-isomer.

3.6.4. (*S*)-2,2,6-Trimethyl-6-vinyltetrahydro-2*H*-pyran (−)-**4**

According to the procedure outlined for the synthesis of ether (+)-**4**, aldehyde (+)-**20** (98% ee) gave, in 70% yield, ether (−)-**4** as a colorless oil with $[\alpha]_D^{20}$ = −9.0 (*c* 2.1, $CHCl_3$) and 95% chemical purity by GC; ^{13}C-NMR, and GC–MS were superimposable to those reported for the (+)-(*R*)-isomer.

Author Contributions: S.S. and D.D.S. contributed equally to the design, and performed the experiments and analyzed the data; S.S. conceived the study and wrote the paper.

Funding: This research was funded by [Regione Lombardia: POR-FESR 2014-2020] grant number [228775 VIPCAT (Value Added Innovative Protocols for Catalytic Transformations)]

Acknowledgments: The authors thank Regione Lombardia for supporting this study within the project POR-FESR 2014-2020 No. 228775 VIPCAT (Value Added Innovative Protocols for Catalytic Transformations).

Conflicts of Interest: The authors declare no conflict of interest.

References

1. Winterhalter, P.; Katzenberger, D.; Schreier, P. 6,7-Epoxy-linalool and related oxygenated terpenoids from *Carica papaya* fruit. *Phytochemistry* **1986**, *25*, 1347–1350. [CrossRef]
2. Avonto, C.; Wang, M.; Chittiboyina, A.G.; Avula, B.; Zhao, J.; Khan, I.A. Hydroxylated bisabolol oxides: Evidence for secondary oxidative metabolism in *Matricaria chamomilla*. *J. Nat. Prod.* **2013**, *76*, 1848–1853. [CrossRef] [PubMed]
3. Boivin, T.L.B. Synthetic routes to tetrahydrofuran, tetrahydropyran, and spiroketal units of polyether antibiotics and a survey of spiroketals of other natural products. *Tetrahedron* **1987**, *43*, 3309–3362. [CrossRef]
4. Nicolaou, K.C.; Prasad, C.V.C.; Somers, P.K.; Hwang, C.K. Activation of 6-endo over 5-exo hydroxy epoxide openings. Stereoselective and ring selective synthesis of tetrahydrofuran and tetrahydropyran systems. *J. Am. Chem. Soc.* **1989**, *111*, 5330–5334. [CrossRef]
5. Nasir, N.M.; Ermanis, K.; Clarke, P.A. Strategies for the construction of tetrahydropyran rings in the synthesis of natural products. *Org. Biomol. Chem.* **2014**, *12*, 3323–3335. [CrossRef] [PubMed]

6. Surburg, H.; Panten, J. *Common Fragrance and Flavor Materials: Preparation, Properties and Uses*, 6th ed.; Wiley-VCH Verlag GmbH & Co., KGaA: Weinheim, Germany, 2016; ISBN 9783527331604.
7. Ohta, Y.; Nishimura, K.; Hirose, Y. Studies on the monoterpene fraction of 'geranium oil' from *Pelargonium roseum Bourbon*. *Agric. Biol. Chem.* **1964**, *28*, 5–9. [CrossRef]
8. Strickler, H.; Kováts, E.S. 245. Zur kenntnis ätherischer öle. Zwei monoterpenoxide aus dem sog. «destillierten» limetten-öl (*Citrus medica* L., *var. acida* Brandis; *Citrus aurantifolia* Swingle). *Helv. Chim. Acta* **1966**, *49*, 2055–2067. [CrossRef]
9. Williams, P.J.; Strauss, C.R.; Wilson, B. Hydroxylated linalool derivatives as precursors of volatile monoterpenes of muscat grapes. *J. Agric. Food Chem.* **1980**, *28*, 766–771. [CrossRef]
10. Engel, K.H.; Tressl, R. Formation of aroma components from nonvolatile precursors in passion fruit. *J. Agric. Food Chem.* **1983**, *31*, 998–1002. [CrossRef]
11. Idstein, H.; Herres, W.; Schreier, P. High-resolution gas chromatography-mass spectrometry and -fourier transform infrared analysis of cherimoya (*Annona cherimolia*, Mill.) volatiles. *J. Agric. Food Chem.* **1984**, *32*, 383–389. [CrossRef]
12. Froehlich, O.; Duque, C.; Schreier, P. Volatile constituents of curuba (*Passiflora mollissima*) fruit. *J. Agric. Food Chem.* **1989**, *37*, 421–425. [CrossRef]
13. Boelens, M.H.; Jimenez, R. Chemical composition of the essential oils from the gum and from various parts of *Pistacia lentiscus* L. (mastic gum tree). *Flavour Frag. J.* **1991**, *6*, 271–275. [CrossRef]
14. Nitz, S.; Kollmannsberger, H. Changes in flavour composition during thermal concentration of apricot purée. *Z. Lebensm. Unters. Forch.* **1993**, *197*, 541–545. [CrossRef]
15. Chen, S.-H.; Huang, T.-C.; Ho, C.-T.; Tsai, P.-J. Extraction, analysis, and study on the volatiles in roselle tea. *J. Agric. Food Chem.* **1998**, *46*, 1101–1105. [CrossRef]
16. Dugo, P.; Cotroneo, A.; Bonaccorsi, I.; Mondello, L. On the genuineness of citrus essential oils. Part LVII. The composition of distilled lime oil. *Flavour Frag. J.* **1998**, *13*, 93–97. [CrossRef]
17. Demyttenaere, J.C.R.; Dagher, C.; Sandra, P.; Kallithraka, S.; Verhé, R.; De Kimpe, N. Flavour analysis of greek white wine by solid-phase microextraction–capillary gas chromatography–mass spectrometry. *J. Chromatogr. A* **2003**, *985*, 233–246. [CrossRef]
18. Quijano, C.E.; Pino, J.A. Analysis of volatile compounds of cacao maraco (*Theobroma bicolor* Humb. et Bonpl.) fruit. *J. Essent. Oil Res.* **2009**, *21*, 211–215. [CrossRef]
19. Hu, C.-D.; Liang, Y.-Z.; Guo, F.-Q.; Li, X.-R.; Wang, W.-P. Determination of essential oil composition from *Osmanthus fragrans* tea by GC-MS combined with a chemometric resolution method. *Molecules* **2010**, *15*, 3683–3693. [CrossRef] [PubMed]
20. Zeng, L.-B.; Zhang, Z.-R.; Luo, Z.-H.; Zhu, J.-X. Antioxidant activity and chemical constituents of essential oil and extracts of *Rhizoma homalomenae*. *Food Chem.* **2011**, *125*, 456–463. [CrossRef]
21. Hamasaki, T.; Sato, Y.; Hatsuda, Y. Isolation of new metabolites from *Aspergillus sydowi* and structure of sydowic acid. *Agric. Biol. Chem.* **1975**, *39*, 2337–2340. [CrossRef]
22. Zdero, C.; Bohlmann, F. Pseudoguaianolides and other constituents from *Anisopappus pinnatifidus* and *Antiphiona* species. *Phytochemistry* **1989**, *28*, 1155–1161. [CrossRef]
23. Nagai, M.; Tanaka, O.; Shibata, S. Chemical studies on the oriental plant drugs. XXVI. Saponins and sapogenins of ginseng. The absolute configurations of cinenic acid and panaxadiol. *Chem. Pharm. Bull.* **1971**, *19*, 2349–2353. [CrossRef]
24. Brenna, E.; Fuganti, C.; Gatti, F.G.; Serra, S. Biocatalytic methods for the synthesis of enantioenriched odor active compounds. *Chem. Rev.* **2011**, *111*, 4036–4072. [CrossRef] [PubMed]
25. Serra, S.; Nobile, I. Chemoenzymatic preparation of the *p*-menth-1,5-dien-9-ol stereoisomers and their use in the enantiospecific synthesis of natural *p*-menthane monoterpenes. *Tetrahedron Asymmetry* **2011**, *22*, 1455–1463. [CrossRef]
26. Serra, S.; Cominetti, A.A. An expedient synthesis of linden ether. *Nat. Prod. Commun.* **2014**, *9*, 293–296. [PubMed]
27. Serra, S. Recent advances in the synthesis of carotenoid-derived flavours and fragrances. *Molecules* **2015**, *20*, 12817–12840. [CrossRef] [PubMed]
28. Serra, S.; De Simeis, D. A study on the lipase-catalysed acylation of 6,7-dihydroxy-linalool. *Nat. Prod. Commun.* **2016**, *11*, 1217–1220.

29. Serra, S.; De Simeis, D.; Brenna, E. Lipase mediated resolution of *cis*- and *trans*-linalool oxide (pyranoid). *J. Mol. Catal. B Enzym.* **2016**, *133*, S420–S425. [CrossRef]

30. Vidari, G.; Giori, A.; Dapiaggi, A.; Lanfranchi, G. Asymmetric dihydroxylation of linalool, nerolidol and citronellyl acetate. Enantioselective synthesis of (3*S*,6*S*)-tetrahydro-2,2,6-trimethyl-6-vinyl-2*H*-pyran-3-ol. *Tetrahedron Lett.* **1993**, *34*, 6925–6928. [CrossRef]

31. Vidari, G.; Lanfranchi, G.; Pazzi, N.; Serra, S. Studies on the total synthesis of the saponaceolides. 1. Enantioselective synthesis of the spiroketal subunit. *Tetrahedron Lett.* **1999**, *40*, 3063–3066. [CrossRef]

32. Aprotosoaie, A.C.; Hăncianu, M.; Costache, I.-I.; Miron, A. Linalool: A review on a key odorant molecule with valuable biological properties. *Flavour Frag. J.* **2014**, *29*, 193–219. [CrossRef]

33. Rupe, H.; Lang, G. 2,6-dimethyl-hexan-äthinoxyd. *Helv. Chim. Acta* **1929**, *12*, 1133–1141. [CrossRef]

34. Serra, S. Bisabolane sesquiterpenes: Synthesis of (*R*)-(+)-sydowic acid and (*R*)-(+)-curcumene ether. *Synlett* **2000**, *2000*, 890–892.

35. Strickler, H.; Ohloff, G. 253. Die chiralität der α-cinensäure. *Helv. Chim. Acta* **1966**, *49*, 2157–2161. [CrossRef]

36. Carlsen, P.H.J.; Katsuki, T.; Martin, V.S.; Sharpless, K.B. A greatly improved procedure for ruthenium tetroxide catalyzed oxidations of organic compounds. *J. Org. Chem.* **1981**, *46*, 3936–3938. [CrossRef]

37. Ishizumi, K.; Koga, K.; Yamada, S. Chemistry of sodium borohydride and diborane. IV. Reduction of carboxylic acids to alcohols with sodium borohydride through mixed carbonic-carboxylic acid anhydrides. *Chem. Pharm. Bull.* **1968**, *16*, 492–497. [CrossRef]

38. Serra, S.; Fuganti, C. Enzyme-mediated preparation of enantiomerically pure *p*-menthan-3,9-diols and their use for the synthesis of natural *p*-menthane lactones and ethers. *Helv. Chim. Acta* **2002**, *85*, 2489–2502. [CrossRef]

39. Serra, S.; Fuganti, C.; Gatti, F.G. A chemoenzymatic, preparative synthesis of the isomeric forms of *p*-menth-1-en-9-ol: Application to the synthesis of the isomeric forms of the cooling agent 1-hydroxy-2,9-cineole. *Eur. J. Org. Chem.* **2008**, *2008*, 1031–1037. [CrossRef]

40. Serra, S. Lipase-mediated resolution of substituted 2-aryl-propanols: Application to the enantioselective synthesis of phenolic sesquiterpenes. *Tetrahedron Asymmetry* **2011**, *22*, 619–628. [CrossRef]

41. Serra, S.; Piccioni, O. A new chemo-enzymatic approach to the stereoselective synthesis of the flavors tetrahydroactinidiolide and dihydroactinidiolide. *Tetrahedron Asymmetry* **2015**, *26*, 584–592. [CrossRef]

42. Chen, C.S.; Fujimoto, Y.; Girdaukas, G.; Sih, C.J. Quantitative analyses of biochemical kinetic resolutions of enantiomers. *J. Am. Chem. Soc.* **1982**, *104*, 7294–7299. [CrossRef]

43. Burdock, G.A. *Fenaroli's Handbook of Flavor Ingredients*, 6th ed.; CRC Press: Boca Raton, FL, USA, 2010; pp. 1957–1958. ISBN 978-1-4200-9077-2.

44. Tschumi, R.; Bonrath, W.; Tschumi, J. Process of Production of Cyclo-Dehydrolinalool. WO Patent 2016/128423 A1, 18 August 2016.

45. Klein, E.; Farnow, H.; Rojahn, W. Die chemie der linalool-oxide. *Liebigs Ann. Chem.* **1964**, *675*, 73–82. [CrossRef]

46. Parikh, J.R.; Doering, W.v.E. Sulfur trioxide in the oxidation of alcohols by dimethyl sulfoxide. *J. Am. Chem. Soc.* **1967**, *89*, 5505–5507. [CrossRef]

47. Takai, K.; Hotta, Y.; Oshima, K.; Nozaki, H. Effective methods of carbonyl methylenation using CH_2I_2-Zn-Me_3Al and CH_2Br_2-Zn-$TiCl_4$ system. *Tetrahedron Lett.* **1978**, *19*, 2417–2420. [CrossRef]

catalysts

MDPI

Article

Biocatalytic Approach to Chiral β-Nitroalcohols by Enantioselective Alcohol Dehydrogenase-Mediated Reduction of α-Nitroketones

Francesca Tentori [1,†], Elisabetta Brenna [1,2,*] (ID), Danilo Colombo [1], Michele Crotti [1], Francesco G. Gatti [1] (ID), Maria Chiara Ghezzi [1] and Giuseppe Pedrocchi-Fantoni [2]

[1] Politecnico di Milano, Dipartimento di Chimica, Materiali e Ingegneria Chimica, Via Mancinelli 7, Milano I-20131, Italy; francesca.tentori@polimi.it (F.T.); danilo.colombo@polimi.it (D.C.); michele.crotti@polimi.it (M.C.); francesco.gatti@polimi.it (F.G.G.); maria.ghezzi@studenti.unimi.it (M.C.G.)

[2] Istituto di Chimica del Riconoscimento Molecolare–CNR, Via Mancinelli 7, Milano I-20131, Italy; giuseppe.pedrocchi@polimi.it

* Correspondence: mariaelisabetta.brenna@polimi.it; Tel.: +39-02-2399-3077

† With the exception of the first author, the others are listed in alphabetical order.

Received: 12 July 2018; Accepted: 25 July 2018; Published: 29 July 2018

check for updates

Abstract: Chiral β-nitroalcohols are important building blocks in organic chemistry. The synthetic approach that is based on the enzyme-mediated reduction of α-nitroketones has been scarcely considered. In this work, the use of commercial alcohol dehydrogenases (ADHs) for the reduction of aromatic and aliphatic nitroketones is investigated. High conversions and enantioselectivities can be achieved with two specific ADHs, affording either the (S) or (R)-enantiomer of the corresponding nitroalcohols. The reaction conditions are carefully tuned to preserve the stability of the reduced product, and to avoid the hydrolytic degradation of the starting substrate. The further manipulation of the enantioenriched nitroalcohols into Boc-protected amminoalcohols is also described.

Keywords: nitroketone; reduction; alcohol-dehydrogenase; enantioselectivity

---

1. Introduction

Chiral β-nitroalcohols **1** (Scheme 1) are relevant synthetic targets in organic chemistry. They are employed as key intermediates for the preparation of a wide range of biologically active natural products and active pharmaceutical ingredients [1–5], especially because they can be readily converted into chiral β-aminoalcohols **2** by reduction of the nitro moiety.

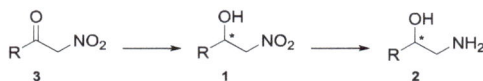

Scheme 1. Synthesis of β-amminoalcohols **2** through β-nitroalcohols **1** as intermediates

The most common approach to compounds **1** is represented by the enantioselective Henry (nitroaldol) reaction between aldehydes and nitromethane, which is catalysed by metal complexes or organocatalysts [1,6–16].

During the past decade, the search for greener and more sustainable synthetic procedures has promoted the investigation of biocatalysed strategies for the synthesis of enantiopure β-nitroalcohols [17]. Several examples of kinetic resolution of racemic compounds **1** catalysed by hydrolases have been reported in the literature [7]. It has also been discovered that some hydroxynitrile lyases (HNLs) are able to promote the enantioselective addition of nitromethane to aldehydes, such as the (S)-selective HNLs from *Hevea brasiliensis* and from *Manihot esculenta* [18–20], and the (R)-selective HNLs from *Arabidopsis thaliana* [21], *Acidobacterium capsulatum*, and *Granulicella tundricula* [22]. These reactions are generally characterized by long reaction times, and strong substrate dependence.

Another possible enzymatic approach, which has received scarce consideration until now, is represented by the bioreduction of α-nitroketones **3**. Only a few papers on this topic are present in the literature. In 1987 [23], the baker's yeast reduction of 3-methyl-3-nitro-2-butanone to the (S)-enantiomer of corresponding alcohol (enantiomeric excess = ee > 96%) in 57% yield was described. A few years later, Moran et al. [24] investigated the reduction of α-nitroacetophenone (**3a**. R = Ph) in fermenting baker's yeast. Only 6% of nitroalcohol **1a** (R = Ph) could be isolated, with benzoic acid being the main product of the biotransformation (27%). According to the authors, the formation of benzoic acid was due to the retro-Henry degradation of nitroalcohol **1a** to benzaldehyde, followed by oxidation. In 2008, Kroutil et al. [25] reported on the conversion of 1-nitro-3-phenylpropan-2-one and 1-nitro-2-octanone into the enantiopure (S)-nitroalcohols in 47% and 75% yield, respectively, by using the lyophilized cells of *Comomonas testoteroni*. Recently [26], the whole cells of *Candida parapsilosis* ATCC 7330 were employed to catalyse the enantioselective reduction of some aliphatic derivatives **3** (only R = alkyl) in water with ethanol as a cosolvent, at room temperature, and in 4 h reaction time (conversion yields 54–76%, ee = 8.2–81%). The formation of the (R) or (S) enantiomer of the corresponding nitroalcohol depended upon the nature of the R group.

The scarcity of experimental data on the bioreduction of α-nitroketones, especially for aromatic derivatives, and the current need for biocatalysed synthesis of chiral building blocks for pharmaceutical applications [27–31] led us to investigate the use of commercial alcohol dehydrogenases for the enantioselective reduction of aryl and alkyl α-nitroketones **3** in controlled reaction conditions. We also studied the further manipulation of specific nitroalcohols **1** to prepare aminoalcohols **2**, which have been already employed as key intermediates for the synthesis of active pharmaceutical ingredients, such as levamisole and (R)-tembamide.

2. Results and Discussion

2.1. Synthesis of Nitroketones 3 and Biocatalysed Reduction to Derivatives 1

Nitroketones **3a–o** (Scheme 2) were synthesized according to the literature by derivatization of the corresponding carboxylic acids with carbonyldiimidazole, followed by reaction with the sodium salt of nitromethane, which was obtained in turn by deprotonation of nitromethane with NaH [32].

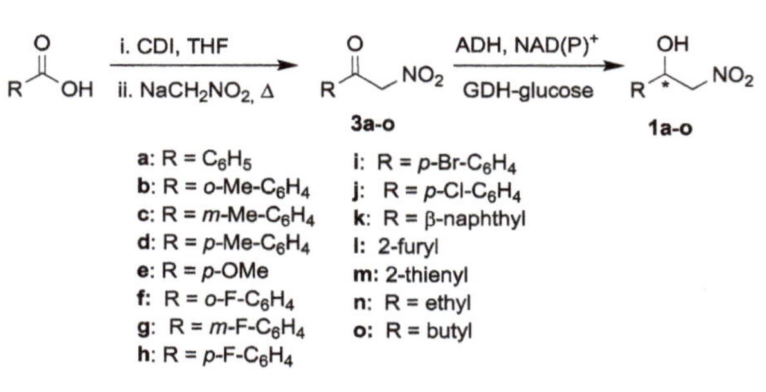

a: R = C₆H₅   i: R = p-Br-C₆H₄
b: R = o-Me-C₆H₄   j: R = p-Cl-C₆H₄
c: R = m-Me-C₆H₄   k: R = β-naphthyl
d: R = p-Me-C₆H₄   l: 2-furyl
e: R = p-OMe   m: 2-thienyl
f: R = o-F-C₆H₄   n: R = ethyl
g: R = m-F-C₆H₄   o: R = butyl
h: R = p-F-C₆H₄

Scheme 2. Synthesis and biocatalysed reduction of nitroketones **3a–o**.

Before starting the alcohol dehydrogenases (ADH) screening, the stability of derivatives **1** was investigated in buffer solutions at pH = 5, 7, and 9 for 4–18 h at 25 °C, using compound **1a** as a model and DMSO as a co-solvent. As expected, nitroalcohol **1a** resulted to be unstable towards retro-Henry reaction in basic and neutral medium: conversion into benzaldehyde was complete at pH = 7 and 9 after 18 h. At pH = 5 no benzaldehyde was observed. Thus, pH = 5 was selected for the investigation of the biocatalysed reduction of compound **3a**, using a panel of 18 commercial alcohol dehydrogenases (from Evoxx). The catalytic NADPH or NADH cofactor was recycled with glucose dehydrogenase (GDH from *Bacillus megaterium*), and glucose was employed as a sacrificial co-substrate. The reactions were performed in acetate buffer solution (pH = 5) with 1% DMSO, monitored by TLC, and usually stopped after 4–5 h. The results of the screening experiments are collected in Table S1. Conversion were evaluated by ^{1}H NMR spectroscopy and the enantiomeric excess values of the reduced products were determined by HPLC analysis on a chiral stationary phase. GC-analysis could not be used because nitroalcohol **1a** undergoes partial thermal degradation to benzaldehyde.

During this screening, benzaldehyde was never detected in the final reaction mixture, while the formation of benzoic acid was observed in a variable amount: from 4–6% in the most effective reductions of **3a** with ADH270 and 440, to nearly 30% in those reactions in which no nitroalcohol was formed. In order to explain the formation of benzoic acid, the stability of compound **3a** was investigated in buffer solution (pH = 5), in the presence of 1% DMSO, GDH, NAD(P)$^+$, without adding the ADH, for 4 and 18 h at 25 °C. Partial degradation (35%) to the carboxylic acid was observed after 4 h, while the complete conversion into benzoic acid was achieved after 18 h. A search in the literature showed that Pearson et al. [33] had described the hydrolytic cleavage of nitroketone **3a** to the corresponding carboxylic acid in dioxane-water solution and the possibility to suppress this side-reaction only in strong mineral acid solution. In the evaluation of the molar percentages of the reduced product **1a** reported in Table S1, as calculated by ^{1}H NMR analysis of the final mixture, the formation of the carboxylic acid was taken into account. The integrals of the following well-separated signals were employed: (i) the doublet of doublets of the CH-OH of **1a** (one hydrogen atom); (ii) the singlet of the CH$_2$ of **3a** (two hydrogen atoms); and, (iii) the doublet of the two aromatic hydrogen atoms adjacent to the COOH group of benzoic acid.

Only eight of the eighteen screened ADHs could catalyze the reduction of nitroketone **3a**. Prolonged reaction times did not improve the yield in the reduction product, instead promoted the extensive hydrolysis of unreacted starting **3a**. The ADHs giving the best results in terms of both conversion and enantioselectivity, i.e., ADH270, 440 and 441, were employed to investigate the reduction of the whole set of nitroketones **3b–o**. The results are reported in Figure 1 and Table S2. The absolute configuration of all the nitroalcohols **1a–o** could be established by a comparison of the corresponding HPLC analyses on chiral stationary phase with those reported in the literature in the same experimental conditions (See Supplementary Materials).

(*R*)-Nitroalcohols were invariably obtained in the presence of ADH440, while opposite enantioselectivity were observed with either ADH270 or ADH441. In the reduction of 2-furyl and 2-thienyl derivatives **3l** and **3m**, obtaining the (*S*)-nitroalcohol with ADH440 and the (*R*)-enantiomer with ADH270, and 441 does not represent an inversion of enantioselectivity with respect to the reductions of the other substrates. It is a consequence of the fact that the priority order of the substituents around the stereogenic centre is different for the presence of the heteroaromatic ring. The only real inversions of configuration were observed in the reduction of **3f** (R = *o*-F-C$_6$H$_4$) and **3n** (R = ethyl) with ADH441 and 270, respectively, affording the corresponding (*R*)-enantiomers with ee = 43 and 80%.

The best results were achieved while using ADH440 as a catalyst (Figure 1). This enzyme promoted the conversion of nitroketones **3** into the (*R*)-enantiomer of nitroalcohols **1** with high yields (c = 79–99%) and very good ee values in the range 92–99% for most of the substrates. Enantioselectivity that was slightly lower 90% was observed in the quantitative reduction of **3c** (R = *m*-Me-C$_6$H$_4$, ee = 84%) and **3l** (R = 2-furyl, ee = 71%). Only in the case of ethyl derivative **3n**, the corresponding reduced product was obtained in racemic form.

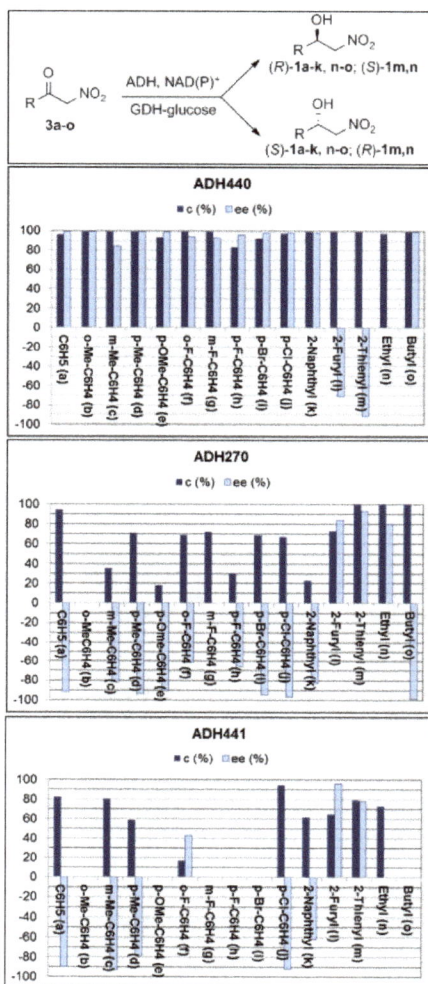

Figure 1. Alcohol dehydrogenases (ADH)-mediated reduction of nitroketones **3a–o** to nitroalcohols **1a–o** (preliminary screening). For graphic reasons the ee values of (*R*)-enantiomers are represented as positive values, those of (*S*)-enantiomers are given as negative: 5 mM substrate, 16 mM glucose, ADH, glucose dehydrogenase (GDH), NAD(P)$^+$, 1% DMSO, acetate buffer pH 5.0, 25 °C, 4–5 h; conversion (c, %) calculated by ^{1}H NMR spectroscopy as molar percentage of the nitroalcohol **1** in the final reaction mixture after 4–5 h, taking into account the unreacted nitroketone **3**, and the carboxylic acid obtained upon nitroketone hydrolysis; enantiomeric excess (ee, %) calculated on the basis of HPLC analysis on a chiral stationary phase.

ADH270 gave the (*S*)-enantiomer of the reduced product in all the bioreductions, with the exception of the reaction of compound **3n** (Figure 1, R = ethyl), affording (*R*)-**1n** (ee = 80%). The highest ee values (91–99%) were obtained in the transformation of *para*-substituted nitroketones **3d** (R = *p*-Me-C$_6$H$_4$, ee = 94%), **3e** (R = *p*-OMe-C$_6$H$_4$, ee = 91%), **3i** (R = *p*-Br-C$_6$H$_4$, ee = 95%), **3j** (R = *p*-Cl-C$_6$H$_4$, ee = 97%), and derivatives **3a** (R = phenyl, ee = 92%), **3m** (R= 2-thienyl, ee = 93%), and **3o** (R = butyl, ee = 99%). Enantioselectivity in the range 80–84% was achieved in the reduction of compounds **3c** (R = *m*-Me-C$_6$H$_4$, ee = 82%), **3k** (R = 2-naphthyl, ee = 83%), **3l** (R = 2-furyl, ee = 84%), and **3n** (R = ethyl, ee = 80%), while modest ee values could be obtained with fluoro derivatives **3f** (R = *o*-F-C$_6$H$_4$, ee = 62%), and **3h** (R = *p*-F-C$_6$H$_4$, ee = 66%). *m*-Fluoro nitroketone **3g** was converted into a racemic nitroalcohol. Only substrate **3b** (R = *o*-Me-C$_6$H$_4$) was recovered unreacted.

When ADH441 was employed as a catalyst (Figure 1), the relevant results were achieved in the reduction of **3a** (R = Ph), **3c** (R = *m*-Me-C$_6$H$_4$), **3j** (R = *p*-Cl-C$_6$H$_4$), and **3l** (R = 2-furyl), affording the corresponding nitroalcohol with high enantiomeric purity (ee = 90, 93, 92 and 96%, respectively).

The results of this screening clearly show that ADH270 and 440 are the most effective catalysts for the preparation of both the enantiomers of nitroalcohols **1**.

2.2. Bioreductions of Nitroketones in Biphasic Medium

In order to avoid the drawback of nitroketone hydrolysis, the use of a biphasic medium (buffer and organic solvent) was evaluated. No benzoic acid was observed when compound **3a** was stirred in toluene/buffer or EtOAc/buffer mixtures for 24 h in the presence of GDH and NAD(P)$^+$ without adding the ADH. In the presence of ADH440 and ADH270 as catalysts, the reductions proceeded affording the results that are reported in Table 1. Toluene resulted to be the solvent of choice, preserving nitroketone **3a** from hydrolysis, still maintaining the activity of the ADH.

Table 1. ADH-mediated reduction of nitroketone **3a** to nitroalcohol **1a** in biphasic system [a].

ADH [1]	Organic Solvent	Conversion [2] (%)	Ee [3] (%)
270	AcOEt	-	-
440	AcOEt	68	98 (*R*)
270	toluene	88	95 (*S*)
440	toluene	99	97 (*R*)

[1] Total volume 4 mL (organic solvent/water 1/1), 6 mM substrate, 20 mM glucose, ADH (2 mg), GDH (1 mg), NAD(P)$^+$ (0.25 mM), acetate buffer pH 5.0, 25 °C, 24 h; [2] conversion calculated on the basis of the ^{1}H NMR spectrum of the crude mixture after 24 h; [3] enantiomeric excess calculated on the basis of HPLC analysis on a chiral stationary phase.

The ADH-mediated reduction of model nitroketone **3a** was also investigated in 1:1 toluene-water (buffer pH = 5) at 25 °C with ADH270 and 440 in order to increase both substrate loading (mg/mL) and substrate to enzyme ratio (mg/mg). The corresponding conversions, determined after 24 h reaction time by ^{1}H NMR spectroscopy, are reported in Table 2.

Table 2. Effect of substrate concentration and substrate/enzyme ratio on conversion for the ADH-mediated reduction of **3a**.

ADH [1]	[Substrate](mg/mL)	Substrate/Enzyme(mg/mg)	Conversion [2](%)
	1	2.0	99
	1	8.0	94
440	2	2.0	95
	2	8.0	91
	3	8.0	94
	3	24.0	80
	1	2.0	88
	1	3.0	67
270	2	2.0	84
	2	3.0	74
	3	3.0	73
	3	4.0	58

[1] Total volume 4 mL (organic solvent/water 1/1), substrate, glucose (3.2 eq), ADH, GDH, NAD(P)$^+$ (0.04 eq), acetate buffer pH 5.0, 25 °C, 24 h; [2] conversion calculated on the basis of the ^{1}H NMR spectrum of the crude mixture after 24 h.

ADH440 was found to be very effective in promoting the reduction of substrate **3a**: in batch conditions, with a substrate concentration of 3 mg/mL, conversion remained still satisfactory (80%) when the enzyme concentration was decreased from 0.38 mg/mL (substrate/enzyme = 8) to 0.12 mg/mL (substrate/enzyme = 24).

ADH 270 showed less efficiency than ADH440 in these bioreductions. When substrate loading was increased to 3 mg/mL the use of 1 mg/mL enzyme (substrate/enzyme = 3) afforded 73% conversion, while a further decrease of enzyme concentration to 0.75 mg/mL (substrate/enzyme = 4) led only to 58% of reduced product.

2.3. Synthesis of Boc-protected β-Aminoalcohols **2**

The conversion of β-nitroalcohols **1** into β-aminoalcohols **2** was investigated, in order to establish the synthetic potential of the ADH-mediated reduction of nitroketones **3** and highlight its value within organic chemistry procedures. Compounds **1a** and **1e** were employed as model substrates, since the corresponding amino derivatives (*S*)-**2a** and (*R*)-**2e** are the key intermediates in the synthesis of active pharmaceutical ingredients, such as levamisole and (*R*)-tembamide (Scheme 3).

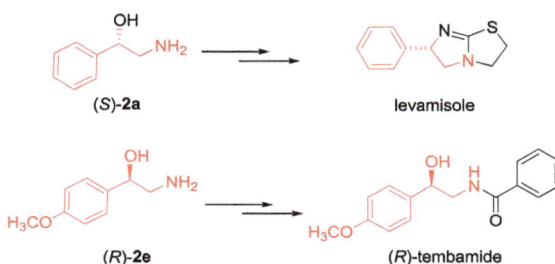

Scheme 3. Active pharmaceutical ingredients prepared starting from amino alcohols (*S*)-**2a** and (*R*)-**2e**.

Levamisole, which is the (*S*)-enantiomer of tetramisole, is a broad spectrum anthelmintic [34], which has found wide application in the treatment of worm infestations and in the elimination of intestinal parasites in both humans and animals. It is also one of the nonspecific immunomodulating agents that are used in clinical practice [35–37]. The known synthetic asymmetric approaches are based on the use of optically active phenylethylenediamine [38–41] or amino alcohol (*S*)-**2a** as intermediates [42].

(*R*)-(−)-Tembamide is a naturally occurring β-hydroxyamide isolated from various members of the Rutaceae family. This compound has been reported to have insecticide and adrenaline-like activity. Extracts of *Aegle marmelos*, containing tembamide, have been used in the Indian traditional medicine as a control for hypoglycemia [43]. Most of the enantioselective procedures to (*R*)-tembamide involve the use of the corresponding amino alcohol (*R*)-**2e** as a key building block [44].

The nitro moiety of model compounds (*S*)-**1a** and (*R*)-**1e** was converted into the corresponding amino functionality by reaction with NiCl$_2$·6H$_2$O and NaBH$_4$ (Scheme 4), followed by treatment with (Boc)$_2$O, in order to facilitate the isolation of the aminoalcohol from the reaction mixture. The reaction was fast and is characterized by complete conversion. The Boc derivatives could be recovered as solid compounds, and easily purified by crystallisation. The procedure was carried out directly in the reaction medium of the biocatalysed reaction, after removal of the aqueous phase, avoiding the isolation of the intermediate nitroalcohols, in order to achieve a one-pot chemo-catalysed conversion of nitroketones **3a** and **3e** into Boc-protected derivatives (*S*)-**2a** and (*R*)-**2e**, in 57 and 63% isolation yields.

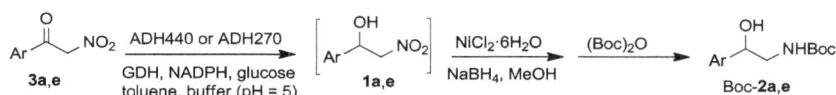

Scheme 4. Synthesis of Boc-protected derivatives **2**.

3. Materials and Methods

3.1. Sources of Enzymes

GDH from *Bacillus megaterium* DSM509 (DSM, Heerlen, the Netherlands) was overexpressed in *E. coli* BL21 (DE3) strains harbouring the plasmid pKTS-GDH prepared according to standard molecular biology techniques. The enzyme was produced and purified, as described in the Supplementary Materials.

The complete set of ADHs was purchased from Evoxx (Monheim am Rhein, Germany).

3.2. General Procedure for the ADH-Mediated Reduction of α-Nitroketones 1a–o (Screening)

A solution of the substrate in DMSO (50 μL, 500 mM) was added to an acetate buffer solution (5 mL, 50 mM, pH 5.0) containing glucose (80 μmol), NADP$^+$ (1 μmol, Sigma-Aldrich, Milan, Italy) or NAD$^+$ (1 μmol, Sigma-Aldrich, Milan, Italy) (according to the ADH preference), GDH (1.5 mg), and the required ADH (3 mg, Evoxx, Monheim am Rhein, Germany). The mixture was incubated for 4–5 h in an orbital shaker (150 rpm, 30 °C). The solution was extracted with EtOAc (2 × 1 mL, Sigma-Aldrich, Milan, Italy), centrifuging after each extraction (15,000 *g*, 1.5 min). The combined organic solutions were dried over anhydrous Na$_2$SO$_4$, and concentrated under reduced pressure. The residue was submitted to ^{1}H NMR analysis (Bruker, Milan, Italy) to determine conversion. Two replicates were performed for each biotransformation: no significant differences (less than 5%) were observed for conversion and enantiomeric excess values.

The enantiomeric excess values of each nitroalcohol was determined by HPLC analysis (Agilent, Cernusco sul Naviglio, Italy) on a chiral stationary phase (See Supplementary Materials). The comparison of these HPLC analyses with those that were reported in the literature in the same experimental conditions (See Supplementary Materials) allowed for the absolute configuration of nitroalcohols **1a–o** to be established.

3.3. General Procedure for the Reduction of Nitroketone 3a in a Biphasic System Mediated by ADH440 and ADH270.

A solution of nitroketone **3a** (4 mg, 25 μmol) in toluene (2 mL) was mixed with an acetate buffer solution (2 mL, 50 mM, pH = 5), containing glucose (80 μmol), NADP$^+$ (1 μmol), GDH (1 mg), and the required ADH (2 mg). The mixture was incubated for 24 h in an orbital shaker (150 rpm, 30 °C). The mixture was extracted with EtOAc (2 × 1 mL), centrifuging after each extraction (15,000 *g*, 1.5 min). The combined organic solutions were dried over anhydrous Na$_2$SO$_4$, and concentrated under reduced pressure. The residue was submitted to ^{1}H NMR analysis to determine conversion. Two replicates were performed for each biotransformation: no significant differences (less than 5%) were observed for the conversion and enantiomeric excess values.

The same procedure was employed to investigate the effect on conversion due to substrate loading and substrate to enzyme ratio, by changing the amount of nitroketone and ADH.

3.4. General Procedure for the Conversion α-Nitroketones 3a and 3e into Boc Protected Amino Alcohols 2a and 2e

The enantioselective reduction of the nitroketone was performed with the required ADH on 100 mg of nitroketone (20 mL of toluene, 20 mL of buffer pH = 5, 15 mg of ADH440 or 35 mg of ADH270, 10 mg NADP$^+$, 7.5 or 18 mg GDH, 150 mg glucose), following the procedure already described in the previous paragraph for biotransformations in biphasic medium. After 24 h, the aqueous phase was removed, methanol was added (0.5 mL), followed by the cautious addition of NiCl$_2$·6H$_2$O (1 eq) and NaBH$_4$ (3 eq) under vigorous stirring. After 30 min, (Boc)$_2$O (1.2 eq) was added. The mixture was stirred for 30 min, filtered through a celite pad, and extracted with EtOAc. The organic layers were dried over anhydrous Na$_2$SO$_4$, and the residue was purified by crystallization from hexane–EtOAc.

3.4.1. (*S*)-*Tert*-butyl (2-hydroxy-2-phenylethyl) carbamate ((*S*)-**2a**)

From compound **3a** (100 mg, 0.61 mmol), using ADH270, derivative (*S*)-**2a** was obtained (83 mg, 57%): ee (HPLC) = 92%, [α]$_D$ = +46.8 (*c* 0.85, CHCl$_3$) [lit. ref. [45] [α]$_D$ = +45.1. (*c* 0.6, CHCl$_3$) for (*S*)-1a with ee = 93%]; ^{1}H NMR (CDCl$_3$, 400 MHz) [39]: δ = 7.40–7.27 (m, 5H, Ar*H*), 4.92 (br s, 1H, N*H*), 4.83 (m, 1H, C*H*OH), 3.48 (m, 1H, C*H*N), 3.26 (m, 1H, C*H*N), 3.01 (br s, 1H, O*H*), 1.45 (s, 9H, (C*H*$_3$)$_3$C); ^{13}C NMR (CDCl$_3$, 100.6 MHz) [45]: δ = 157.1, 142.0, 128.6, 127.9, 126.0, 80.0, 74.0, 48.5, 28.5; GC-MS (EI) t$_r$= 21.5 min *m*/*z* (%) = 181 (M$^+$-56, 14), 107 (100), 79 (47), 57 (100).

HPLC analysis [45]: Chiralcel OD, 95/5 hexane/*i*-PrOH, 0.6 mL/min, 215 nm, (*R*)-**2a** t$_r$ = 19.1 min, (*S*)-**2a** t$_r$ = 23.4 min.

3.4.2. (*R*)-*Tert*-butyl (2-hydroxy-2-(4-methoxyphenyl)ethyl) carbamate ((*R*)-**2e**)

From compound **3e** (100 mg, 0.51mmol), using ADH440, derivative (*R*)-**2e** was obtained (86.3 g, 63%): ee (HPLC) = 96%, [α]$_D$ = −37.7 (c 0.7, CHCl$_3$); ^{1}H NMR (CDCl$_3$, 400 MHz) [46]: δ = 7.28 (d, *J* = 8.7 Hz, 2H, Ar*H*), 6.88 (d, *J* = 8.7 Hz, 2H, Ar*H*), 4.95 (br s, 1H, N*H*), 4.76 (m, 1H, C*H*OH), 3.80 (s, 3H, OC*H*$_3$), 3.43 (m, 1H, C*H*N), 3.23 (m, 1H, C*H*N), 3.00 (br s, 1H, O*H*), 1.44 (s, 9H, (C*H*$_3$)$_3$C); ^{13}C NMR (CDCl$_3$, 100.6 MHz): δ = 159.4, 157.0, 134.1, 127.2, 114.1, 79.9, 73.2, 55.4, 48.5, 28.5; GC-MS (EI) t$_r$= 23.8 min m/z (%) = 267 (M$^+$, 0.5), 211 (5), 137 (100), 109 (15), 57 (18).

HPLC analysis: Chiralcel OD, 95/5 hexane/*i*-PrOH, 0.6 mL/min, 215 nm, (*R*)-**2a** t$_r$ = 26.9 min, (*S*)-**2a** t$_r$ = 34.9 min.

4. Conclusions

The results that are reported in this work show that the biocatalytic reduction of α-nitroketones **3** mediated by ADHs is a convenient and useful procedure for the synthesis of both the enantiomers of the corresponding β-nitroalcohols **1** with high enantiomeric purity. In particular, for the first time the reduction of aryl and heteroaryl α-nitroketones (R = aryl or heteroaryl) has been successfully achieved by enzymatic catalysis, enlarging the known methods for the reduction of these compounds limited up to now to the asymmetric transfer hydrogenation in the presence of ruthenium [47], and iridium [48] chiral complexes, with formic acid as a reductant.

The bioreduction is performed under mild conditions (ambient temperature and pressure), with low energy consumption, at the expense of glucose, which is employed as a sacrificial substrate for the enzymatic regeneration of the cofactor. The enzymes catalyzing this transformation with either (*R*)- and (*S*)-selectivity are commercially available, and they can be manipulated easily and safely. The use of a biphasic reaction medium with toluene as an organic solvent does not inhibit the activity of the selected ADHs, helps in preserving the starting substrate from hydrolytic degradation, and it improves work-up procedures. The further manipulation of nitroketones into aminoalcohols was carried out without isolation of the nitroalcohol intermediate, thus telescoping the synthetic sequence.

Future work will be devoted to increase the productivity of the reaction, for example, by immobilizing the most suitable ADHs on solid supports and performing the reaction in flow conditions.

Supplementary Materials: The following are available online at http://www.mdpi.com/2073-4344/8/8/308/s1.

Author Contributions: E.B. conceived and designed the experiments; F.T., D.C. and M.C.G. performed the experiments and the structural characterization of compounds; E.B., M. C., F.G.G., and G.P.-F. analyzed the data and wrote the paper.

Funding: The authors thank Regione Lombardia for the financial support to the fellowship of D.C. within the project VIPCAT (Value Added Innovative Protocols for Catalytic Transformations"–ID 228775).

Acknowledgments: To Professor Claudio Fuganti with gratitude.

Conflicts of Interest: The authors declare no conflict of interest.

References

1. Luzzio, F.A. The Henry reaction: recent examples. *Tetrahedron* **2001**, *57*, 915–945. [CrossRef]
2. Klingler, F.D. Asymmetric hydrogenation of prochiral amino ketones to amino alcohols for pharmaceutical use. *Acc. Chem. Res.* **2007**, *40*, 1367–1376. [CrossRef] [PubMed]
3. Takasu, T.; Ukai, M.; Sato, S.; Matsui, T.; Nagase, I.; Maruyama, T.; Sasamata, M.; Miyata, K.; Uchida, H.; Yamaguchi, O. Effect of (*R*)-2-(2-aminothiazol-4-yl)-4-{2-[(2-hydroxy-2-phenylethyl)amino]ethyl} acetanilide (YM178), a novel selective beta(3)-adrenoceptor agonist, on bladder function. *J. Pharmacol. Exp. Ther.* **2007**, *321*, 642–647. [CrossRef] [PubMed]
4. Hicks, A.; McCafferty, G.P.; Riedel, E.; Aiyar, N.; Pullen, M.; Evans, C.; Luce, T.D.; Coatney, R.W.; Rivera, G.C.; Westfall, T.D.; et al. GW427353 (solabegron), a novel, selective beta(3)-adrenergic receptor agonist, evokes bladder relaxation and increases micturition reflex threshold in the dog. *J. Pharmacol. Exp. Ther.* **2007**, *323*, 202–209. [CrossRef] [PubMed]
5. Blay, G.; Hernández-Olmos, V.; Pedro, J.P. Synthesis of (*S*)-(+)-sotalol and (*R*)-(−)-isoproterenol via a catalytic enantioselective Henry reaction. *Tetrahedron* **2010**, *21*, 578–581. [CrossRef]
6. Alvarez-Casao, Y.; Marques-Lopez, E.; Herrera, R.P. Organocatalytic enantioselective Henry reactions. *Symmetry* **2011**, *3*, 220–245. [CrossRef]
7. Palomo, C.; Oiarbide, M.; Laso, A. Recent advances in the catalytic asymmetric nitroaldol (Henry) reaction. *Eur. J. Org. Chem.* **2007**, *2561–2574*, 2561–2574. [CrossRef]
8. Marcelli, T.; van der Haas, R.N.S.; van Maarseveen, J.H.; Hiemstra, H. Asymmetric organocatalytic Henry reaction. *Angew. Chem. Int. Ed.* **2006**, *45*, 929–931. [CrossRef] [PubMed]
9. Shibazaki, M.; Kumagaya, N.; Nitabara, T. Preparation of Optically Active Anti-1,2-Nitroalcohols by Stereoselective Nitroaldol Reaction. Jpn. Kokai Tokkyo Koho JP 2009114071 A, 28 May 2009.
10. Shibazaki, M.; Matsunaga, S.; Handa, S. Preparation of Optically Active Anti-1,2-Nitroalcohols by Stereoselective Nitroaldol Reaction with Palladium Lanthanum-Schiff Base Complex Catalysts. Jpn. Kokai Tokkyo Koho JP 2009108012 A, 21 May 2009.
11. Ooi, T.; Uraguchi, D. Chiral Tetraaminophosphonium salts, catalyst for asymmetric synthesis and method for producing chiral beta Nitroalcohol. US 20090131716 A1, 21 May 2009.
12. Baxter, C.E.J. Preparation of (1R*,2S*)-1-Phenyl-2-Nitroalcohols from Benzaldehyde and the Corresponding Nitroalkane in the Presence of Amine Catalysts. US 5750802 A, 12 May 1998.
13. Kodama, K.; Sugawara, K.; Hirose, T. Synthesis of chiral 1,3-diamines derived from *cis*-2-benzamidocyclohexane carboxylic acid and their application in the Cu-catalyzed enantioselective Henry reaction. *Chem. Eur. J.* **2011**, *17*, 13584–13592. [CrossRef] [PubMed]
14. Chinnaraja, E.; Arunachalam, R.; Subramanian, P.S. Enantio- and diastereoselective synthesis of beta-nitroalcohol via Henry reaction catalyzed by Cu(II), Ni(II), Zn(II) complexes of chiral BINIM ligands. *ChemistrySelect* **2016**, *1*, 5331–5338. [CrossRef]
15. Cwiek, R.; Niedziejko, P.; Kaluzia, Z. Synthesis of tunable diamine ligands with spiro indane-2,2-pyrrolidine backbone and their applications in enantioselective Henry reaction. *J. Org. Chem.* **2014**, *79*, 1222–1234. [CrossRef] [PubMed]
16. Angelin, M.; Vongvilai, P.; Fischer, A.; Ramström, O. Crystallization-driven asymmetric synthesis of pyridine-beta-nitroalcohols via discovery-oriented self-resolution of a dynamic system. *Eur. J. Org. Chem.* **2010**, *2010*, 6315–6318. [CrossRef]
17. Milner, S.E.; Moody, T.S.; Maguire, A.R. Biocatalytic approaches to the Henry (nitroaldol) reaction. *Eur. J. Org. Chem.* **2012**, *2012*, 3059–3067. [CrossRef]
18. Purkarthofer, T.; Gruber, K.; Gruber-Khadjawi, M.; Waich, K.; Skranc, W.; Mink, D.; Griengl, H. A biocatalytic henry reaction—the hydroxynitrile lyase from *Hevea brasiliensis* also catalyzes nitroaldol. *Angew. Chem. Int. Ed.* **2006**, *45*, 3454–3456. [CrossRef] [PubMed]

19. Gruber-Khadjawi, M.; Purkarthofer, T.; Skranc, W.; Griengl, H. A biocatalytic Henry reaction—The hydroxynitrile lyase from *Hevea brasiliensis* also catalyzes nitroaldol reactions. *Adv. Synth. Catal.* **2007**, *349*, 1445–1450. [CrossRef]

20. Yuryev, R.; Purkarthofer, T.; Gruber, M.; Griengl, H.; Liese, A. Kinetic studies of the asymmetric Henry reaction catalyzed by hydroxynitrile lyase from *Hevea brasiliensis*. *Biocatal. Biotransform.* **2010**, *28*, 348–356. [CrossRef]

21. Fuhshuku, K.; Asano, Y. Synthesis of (*R*)-beta-nitro alcohols catalyzed by *R*-selective hydroxynitrile lyase from *Arabidopsis thaliana* in the aqueous-organic biphasic system. *J. Biotechnol.* **2011**, *153*, 153–159. [CrossRef] [PubMed]

22. Bekerle-Bogner, M.; Gruber-Khadjawi, M.; Wiltsche, H.; Wiedner, R.; Schwab, H.; Steiner, K. (*R*)-Selective nitroaldol reaction catalyzed by metal-dependent bacterial hydroxynitrile lyase. *ChemCatChem* **2016**, *8*, 2214–2216. [CrossRef]

23. Fujisawa, T.; Hayashi, H.; Kishioka, Y. Enantioselective Synthesis of Optically Pure Amino Alcohol Derivatives by Yeast Reduction. *Chem. Lett.* **1987**, *16*, 129–132. [CrossRef]

24. Brenelli, E.; de Carvalho, M.; Marques, M.; Moran, P.J.S.; Rodrigues, J.A.R.; Sorrilha, A.E.P.M. Enantioselective synthesis of (*R*)-(−)-phenylethanolamines using Baker's yeast reduction of some substituted methyl phenyl ketones. *Indian J. Chem.* **1992**, *31B*, 821–823.

25. Wallner, S.R.; Lavandera, I.; Mayer, S.F.; Öhrlein, R.; Hafner, A.; Edegger, K.; Faber, K.; Kroutil, W. Stereoselective anti-Prelog reduction of ketones by whole cells of Comamonas testosteroni in a substrate-coupled approach. *J. Mol. Catal. B* **2008**, *55*, 126–129. [CrossRef]

26. Venkataraman, S.; Chadha, A. Enantio- & chemo-selective preparation of enantiomerically enriched aliphatic nitro alcohols using Candida parapsilosis ATCC 7330. *RSC Adv.* **2015**, *5*, 73807–73813. [CrossRef]

27. Albarrán-Velo, J.; González-Martínez, D.; Gotor-Fernández, V. Stereoselective biocatalysis: A mature technology for the asymmetric synthesis of pharmaceutical building blocks. *Biocatal. Biotransform.* **2017**, *36*, 102–130. [CrossRef]

28. Brenna, E.; Crotti, M.; Gatti, F.G.; Monti, D.; Parmeggiani, F.; Pugliese, A.; Tentori, F. Biocatalytic synthesis of chiral cyclic gamma-oxoesters by sequential C−H hydroxylation, alcohol oxidation and alkene reduction. *Green Chem.* **2017**, *19*, 5122–5130. [CrossRef]

29. Brenna, E.; Crotti, M.; Gatti, F.G.; Monti, D.; Parmeggiani, F.; Santangelo, S. Asymmetric bioreduction of beta-acylaminonitroalkenes: Easy access to chiral building blocks with two vicinal nitrogen-containing functional groups. *ChemCatChem* **2017**, *9*, 2480–2487. [CrossRef]

30. Brenna, E.; Cannavale, F.; Crotti, M.; De Vitis, V.; Gatti, F.G.; Migliazza, G.; Molinari, F.; Parmeggiani, F.; Romano, D.; Santangelo, S. Synthesis of enantiomerically enriched 2-hydroxymethylalkanoic acids by oxidative desymmetrisation of achiral 1,3-diols mediated by *Acetobacter aceti*. *ChemCatChem* **2016**, *8*, 3796–3803. [CrossRef]

31. Brenna, E.; Crotti, M.; Gatti, F.G.; Monti, D.; Parmeggiani, F.; Powell, R.W.; Santangelo, S.; Stewart, J.D. Opposite enantioselectivity in the bioreduction of (*Z*)-beta-aryl-beta-cyanoacrylates mediated by the tryptophan 116 mutants of Old Yellow Enzyme 1: synthetic approach to (*R*)- and (*S*)-beta-aryl-gamma-lactams. *Adv. Synth. Catal.* **2015**, *357*, 1849–1860. [CrossRef]

32. Lindsay, V.N.G.; Lin, W.; Charette, A.B. Experimental evidence for the all-up reactive conformation of chiral rhodium(II) carboxylate catalysts: enantioselective synthesis of *cis*-cyclopropane alpha-amino acids. *J. Am. Chem. Soc.* **2009**, *131*, 16383–16385. [CrossRef] [PubMed]

33. Pearson, R.G.; Anderson, D.H.; Alt, L.L. Mechanism of the hydrolytic cleavage of carbon-carbon bonds. III. Hydrolysis of alpha-nitro and alpha-sulfonyl ketones. *J. Am. Chem. Soc.* **1955**, *77*, 527–529. [CrossRef]

34. Thienpont, D.C.I.; Vanparijs, O.; Raeymaekers, A.; Vandenberk, J.; Demoen, P.; Allewijn, F.; Marsboom, R.; Niemegeers, C.; Schellekens, K.; Janssen, P. Tetramisole (R 8299), a new, potent broad spectrum anthelmintic. *Nature* **1966**, *209*, 1084–1086. [CrossRef] [PubMed]

35. Moertel, C.G.; Fleming, T.T.; Macdonald, J.S.; Haller, D.G.; Laurie, J.A.; Tangen, C.M.; Ungerleider, J.S.; Emerson, W.A.; Tormey, D.C.; Glick, J.H.; et al. Fluorouracil plus levamisole as effective adjuvant therapy after resection of stage III colon carcinoma: a final report. *Anal. Int. Med.* **1995**, *122*, 321–326. [CrossRef]

36. Moertal, G.G.; Fleming, T.R.; Macdonald, J.S. Levamisole and fluorouracil for adjuvant therapy of resected colon-carcinoma. *New Engl. J. Med.* **1990**, *322*, 352–358. [CrossRef] [PubMed]

37. Gwilt, P.; Tempero, M.; Kremer, A.; Connolly, M.; Ding, C. Pharmacokinetics of levamisole in cancer patients treated with 5-fluorouracil. *Cancer Chemother. Pharmacol.* **2000**, *45*, 247–251. [CrossRef] [PubMed]

38. Raeymaekers, A.H.M.; Roevens, L.F.C.; Janssen, P.A.J. The absolute configurations of the optical isomers of the broad spectrum anthelmintic tetramisole. *Tetrahedron Lett.* **1967**, *8*, 1467–1470. [CrossRef]

39. Choudhary, M.K.; Rajkumar Tak, R.; Kureshy, R.I.; Ansari, A.; Khan, N.H.; Abdi, S.H.R.; Bajaj, H.C. Enantioselective aza-Henry reaction for the synthesis of (S)-levamisole using efficient recyclable chiral Cu(II)-amino alcohol derived complexes. *J. Mol. Catal. A.* **2015**, *409*, 85–93. [CrossRef]

40. Roeben, C.; Souto, J.A.; Gonzalez, Y.; Lishchynskyi, A.; Muniz, K. Enantioselective metal-free diamination of styrenes. *Angew. Chem., Int. Ed.* **2011**, *50*, 9478–9482. [CrossRef] [PubMed]

41. Sadhukhan, A.; Sahu, D.; Ganguly, B.; Khan, N.H.; Kureshy, R.I.; Abdi, S.H.R.S.; Bajaj, H.C. Oxazoline-based organocatalyst for enantioselective Strecker reactions: A protocol for the synthesis of levamisole. *Chem. Eur. J.* **2013**, *19*, 14224–14232. [CrossRef] [PubMed]

42. Kamal, A.; Ramesh Khanna, G.B.; Krishnaji, T.; Ramu, R. A new facile chemoenzymatic synthesis of levamisole. *Biorg. Med. Chem. Lett.* **2005**, *15*, 613–615. [CrossRef] [PubMed]

43. Shoeb, A.; Kapil, R.S.; Popli, S.P. Coumarins and alkaloids of Aegle-marmelos. *Phytochemistry* **1973**, *12*, 2071–2072. [CrossRef]

44. Das, A.; Choudhary, M.K.; Kureshy, R.I.; Roy, T.; Khan, N.H.; Abdi, S.H.R.; Bajaj, H.C. Enantioselective Henry and aza-Henry reaction in the synthesis of (R)-tembamide using efficient, recyclable polymeric CuII complexes as catalyst. *ChemPlusChem* **2014**, *79*, 1138–1146. [CrossRef]

45. Russo, A.; Lattanzi, A. Catalytic asymmetric beta-peroxidation of nitroalkenes. *Adv. Synth. Catal.* **2008**, *350*, 1991–1995. [CrossRef]

46. O'Brien, P.; Osborne, S.A.; Parker, D.D. Asymmetric aminohydroxylation of substituted styrenes: applications in the synthesis of enantiomerically enriched arylglycinols and a diamine. *J. Chem. Soc. Perkin Trans.* **1998**, *1*, 2519–2526. [CrossRef]

47. Watanabe, M.; Murata, K.; Ikariya, T. Practical synthesis of optically active amino alcohols via asymmetric transfer hydrogenation of functionalized aromatic ketones. *J. Org. Chem.* **2002**, *67*, 1712–1715. [CrossRef] [PubMed]

48. Soltani, O.; Ariger, M.A.; Vázquez-Villa, H.E.; Carreira, M. Transfer hydrogenation in water: enantioselective, catalytic reduction of alpha-cyano and alpha-nitro substituted acetophenones. *Org. Lett.* **2010**, *12*, 2893–2895. [CrossRef] [PubMed]

catalysts

Article

Stereoselective Enzymatic Reduction of 1,4-Diaryl-1,4-Diones to the Corresponding Diols Employing Alcohol Dehydrogenases

Ángela Mourelle-Insua [1], Gonzalo de Gonzalo [2,*] ORCID, Iván Lavandera [1,*] ORCID and Vicente Gotor-Fernández [1,*] ORCID

[1] Organic and Inorganic Chemistry Department, University of Oviedo, Avenida Julián Clavería 8, 33006 Oviedo, Spain; a.mourelle-insua@outlook.com

[2] Departamento de Química Orgánica, Universidad de Sevilla, c/Profesor García González, 41012 Sevilla, Spain

* Correspondence: gdegonzalo@us.es (G.d.G.); lavanderaivan@uniovi.es (I.L.); vicgotfer@uniovi.es (V.G.-F.); Tel.: +34-954-59997 (G.d.G.); +34-985-103452 (I.L.); +34-985-103454 (V.G.-F.)

check for updates

Received: 18 January 2018; Accepted: 3 April 2018; Published: 6 April 2018

Abstract: Due to the steric hindrance of the starting prochiral ketones, the preparation of chiral 1,4-diaryl-1,4-diols through the asymmetric hydrogen transfer reaction has been mainly restricted to the use of metal-based catalysts, oxazaborolidines, or organocatalysts. Herein, we demonstrated the versatility of oxidoreductases, finding overexpressed alcohol dehydrogenase from *Ralstonia* sp. (*E. coli*/RasADH) as the most active and stereoselective biocatalyst. Thus, the preparation of a set of 1,4-diaryl-1,4-diols bearing different pattern substitutions in the aromatic ring was achieved with complete diastereo- and enantioselectivity under mild reaction conditions.

Keywords: alcohol dehydrogenases; asymmetric synthesis; bioreduction; 1,4-diols; diketones

1. Introduction

Optically active 1,4-diols can be found as structural motifs in a huge number of biologically active compounds as pharmaceuticals, flavors, and fragrances, but also in chiral ligands and auxiliaries for asymmetric synthesis purposes [1,2]. Their synthetic versatility has also been utilized as precursors of valuable compounds such as 2,5-disubstituted pyrrolidines and phosphine derivatives used as ligands in asymmetric hydrogenations [3–5]. In this context, 1,4-butanediol plays a key role due its applications in the food and cosmetic industry, but also as precursor of plastics, pharmaceuticals, fibers, solvents, and biologically active lactones, among others. For these reasons, the preparation of enantioenriched 1,4-diols using efficient and selective methodologies is a field of interest in organic chemistry. Different asymmetric strategies have been reported for their synthesis, the catalytic asymmetric transfer hydrogenation (CATH) of prochiral 1,4-dicarbonylic compounds being one of the most recurrent methods [6,7]. CATH is the most straightforward approach for obtaining the optically pure diols in quantitative yield, but presents difficulty in that the hydroxy ketone intermediate can also be attained and that (up to) four possible diastereoisomers can be formed. Therefore, several selective chiral catalysts and reducing reagents have been developed for the stereoselective asymmetric reduction of these prochiral diketones, including rhodium or iridium complexes [8], different borane complexes such as oxazaborolidines [9,10], or proline-type derivatives as catalysts in combination with the appropriate reducing agents [11,12].

Biocatalysis has emerged in the last decades as a mature technology for the production of optically active (poly)alcohols by means of the stereoselection displayed by different enzymes [13–17].

Among the wide set of (poly)alcohol structures that can be obtained using biocatalysts, the production of optically active 1,4-dialkyl-1,4-diols and their corresponding acetates has been described mainly by the action of lipases through resolution procedures involving acylation and transesterification processes [18–25], and only the resolution of the bulkier 1-phenylpentane-1,4-diol or its acetate has been successfully achieved [26,27]. Alcohol dehydrogenases (ADHs) are valuable biocatalysts that have been widely employed for the synthesis of chiral alcohols due to their usually excellent selectivity [28–31], including dicarbonyl bioreduction processes [32,33]. Among the wide set of alcohol structures that can be obtained with these enzymes, optically active 1,4-dialkyl-1,4-diols have been described. Efforts have been especially focused on the preparation of chiral 2,5-hexanediol, a key starting material for the preparation of several catalysts and drugs [34–37]. However, until now there are no examples about the enzymatic reduction of 1,4-diaryl-1,4-diketones to obtain the corresponding chiral 1,4-diaryl-1,4-diols, as these substrates are probably too bulky for ADH recognition. In the present paper, we describe for the first time the use of ADHs for the stereoselective preparation under mild reaction conditions (aqueous buffer and 30 °C) of a representative panel of 1,4-diaryl-1,4-diols bearing different pattern substitutions in the aromatic ring.

2. Results and Discussion

As a first step, the chemical synthesis of prochiral 1,4-diketones **1–8a** was developed to later study the biocatalyzed stereoselective synthesis of the optically active 1,4-diols **1–8b**. Different methods have been described for the preparation of these substrates [38,39]. Among them, we have carried out the synthesis of these compounds through the cross aldol condensation of 4′-substituted acetophenones with α-bromo-4′-substituted acetophenones, followed by 1,3-dehydrobromination of these products, and cleavage of the corresponding activated cyclopropyl intermediates in the presence of $ZnCl_2$, ^tBuOH, and Et_2NH (Scheme 1) [40]. Thus, 1,4-diketones **1–8a** were obtained in yields varying from 8% for the *o*-methoxy derivative (**8a**) to 58% for 1,4-diphenylbutane-1,4-dione (**1a**).

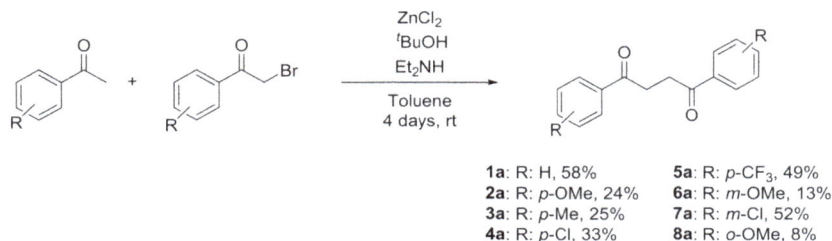

1a: R: H, 58%   **5a**: R: *p*-CF$_3$, 49%
2a: R: *p*-OMe, 24%   **6a**: R: *m*-OMe, 13%
3a: R: *p*-Me, 25%   **7a**: R: *m*-Cl, 52%
4a: R: *p*-Cl, 33%   **8a**: R: *o*-OMe, 8%

Scheme 1. Synthesis of diketones **1–8a** from the corresponding 4′-substituted acetophenones and the α-bromo-4′-substituted acetophenones.

Once synthesized, the obtained diketones were tested in the bioreduction processes catalyzed by a set of commercially available and "made in house" alcohol dehydrogenases. As model substrate 1,4-diphenylbutane-1,4-dione (**1a**) was chosen. Thus, lyophilized cells of *E. coli* overexpressing alcohol dehydrogenases from *Ralstonia* sp. (*E. coli*/RasADH) [41], *Lactobacillus brevis* (*E. coli*/LBADH) [42], *Sphingobium yanoikuyae* (*E. coli*/SyADH) [43], *Thermoanaerobacter ethanolicus* (*E. coli*/TeSADH) [44] and *Rhodococcus ruber* (*E. coli*/ADH-A) [45] were tested. From previous studies, 25 mM substrate concentration was initially considered and 2.5% *v/v* of dimethylsulfoxide (DMSO) was used as cosolvent due to the low solubility of the ketones in the aqueous media, using Tris·HCl buffer 50 mM pH 7.5, and isopropanol (IPA) or glucose dehydrogenase (GDH) with glucose as reducing agents to recycle the nicotinamide cofactor [46]. The reactions were incubated for 24 h at 30 °C, observing only activity when *E. coli*/RasADH was utilized (Table 1). This result is not entirely surprising, as this NADP-dependent enzyme has been previously described as a valuable biocatalyst for the reduction of sterically hindered substrates [47]. Hence, a conversion of 56% was reached by using this enzyme, leading to the formation of diol (1*S*,4*S*)-**1b** in a highly selective manner (98% *de*, >99% *ee*), not observing the hydroxy ketone intermediate (Table 1, entry 1). Remarkably, 82% conversion was obtained after 48 h maintaining the excellent selectivity (Table 1, entry 2).

In order to optimize the bioreduction using this enzymatic preparation, different conditions were studied (Table 1, entries 3–17). Firstly, in similar conditions the absence of cosolvent led to a decrease in the conversion (72%–82% conversion, entries 2–3). When the usual regeneration system (glucose/GDH) was changed to IPA (entry 4), the diastereoselectivity of the process dropped from 98% *de* to 78% *de*.

Table 1. Bioreduction of 1,4-diphenylbutane-1,4-dione (**1a**) using *E. coli*/RasADH [a].

Entry	Regeneration System	Cosolvent	% (v/v)	t (h)	Conversion (%) [b]	de [b,c]	ee (%) [b,d]
1	Glucose/GDH	DMSO	2.5	24	56	99:1	>99
2	Glucose/GDH	DMSO	2.5	48	82	99:1	>99
3	Glucose/GDH	-	-	48	72	99:1	>99
4	IPA	-	-	48	72	89:11	>99
5	Glucose/GDH	EtOH	2.5	24	57	99:1	>99
6	Glucose/GDH	1,4-dioxane	2.5	24	72	99:1	>99
7	Glucose/GDH	1,4-dioxane	5	24	78	>99	>99
8	Glucose/GDH	1,4-dioxane	10	24	79	99:1	>99
9	Glucose/GDH	1,4-dioxane	10	48	80	99:1	>99
10	Glucose/GDH	MTBE	2.5	24	70	>99	>99
11	Glucose/GDH	MTBE	5	24	83	>99	>99
12	Glucose/GDH	MTBE	10	24	77	>99	>99
13	Glucose/GDH	MTBE	10	48	77	>99	>99
14	Glucose/GDH	THF	2.5	24	72	99:1	>99
15	Glucose/GDH	THF	5	24	88	>99	>99
16	Glucose/GDH	THF	10	24	80	99:1	>99
17	Glucose/GDH	THF	10	48	82	99:1	>99

[a] For reaction details, see Materials and methods section; [b] Measured by High Performance Liquid Chromatography (HPLC); [c] Ratio of (1*S*,4*S*) and (1*R*,4*R*) to *meso*-diol; [d] Enantiomeric excess of (1*S*,4*S*)-**1b**.

Taking all these data into account, an exhaustive cosolvent screening was performed (entries 5–17). Ethanol (EtOH), 1,4-dioxane, methyl *tert*-butyl ether (MTBE), and tetrahydrofuran (THF) were chosen as suitable cosolvents for the biotransformation and employed in different quantities. When using 2.5% *v/v* of the organic solvent, for instance 1,4-dioxane, MTBE, and THF (entries 6, 10, and 14), better conversions than DMSO (entry 1) and EtOH (entry 5) were achieved, also affording excellent selectivities. An increase in the amount of these cosolvents from 2.5% to 5% *v/v* revealed even better conversions (entries 7, 11 and 15) and perfect diastereoselectivities (>99% *de*). Finally, the amount of cosolvent was increased up to 10% *v/v* (entries 8, 12, and 16), but unfortunately this did not lead to any improvement. Moreover, higher reaction times (from 24 h to 48 h, entries 9, 13, and 17) did not show better results. It must be pointed out that the hydroxy ketone intermediate was not observed in any case.

Commercial ADHs purchased from Codexis were also tested, as these enzymes have recently demonstrated interesting applications [48]. In this case, IPA was utilized as hydrogen source and phosphate buffer 125 mM pH 7.0 as suitable reaction medium. DMSO (2.5% *v/v*) was added again to solubilize the ketone. Thus, diketone **1a** was a suitable substrate for five of them (69–89% conversion, 74–99% *de*, Table 2). ADH-P2-D03 led to the formation of the opposite enantiomer (1*R*,4*R*)-**1b** (entry 4) with excellent enantioselectivity and 74% *de*. Thus, by modifying the biocatalyst, both enantiomers of diol **1b** could be achieved. Again, the hydroxy ketone intermediate was not detected.

Table 2. Bioreduction of 1,4-diphenylbutane-1,4-dione (**1a**) using commercially available alcohol dehydrogenases (ADHs) [a].

Entry	ADH	Conversion (%) [b]	*de* [b,c]	*ee* (%) [b,d]
1	P1-B02	85	>99	>99
2	P1-B10	79	>99	>99
3	P1-B12	75	99:1	>99
4	P2-D03	89	87:13	>99 [e]
5	P2-D11	69	98:2	>99

[a] For reaction conditions, see Materials and methods section; [b] Measured by HPLC; [c] Ratio of (1*S*,4*S*) and (1*R*,4*R*) to *meso*-(1*R*,4*S*); [d] Enantiomeric excess of (1*S*,4*S*)-**1b**; [e] Enantiomeric excess of (1*R*,4*R*)-**1b**.

Seeking further exploitation of the synthetic approach, first we considered the study of similar substrates that differed at the *para* substitution of the aromatic ring. For this purpose, biotransformations with *E. coli*/RasADH and substrates **2–5a** were set up using the best conditions found in the reduction of diketone **1a** (Table 1, entry 15, 5% *v*/*v* THF, 30 °C, 24 h). The results are summarized in Scheme 2. In all cases, sole formation of (1*S*,4*S*)-diols **2–5b** was observed (>99% *de* and >99% *ee*), with no detection of the corresponding hydroxy ketone intermediates in any case. For chlorinated substrate **4a** (82% conversion), the results were comparable to those obtained with the model substrate (88% conversion). A small drop in the conversion was observed for methoxylated (**2a**) and methylated (**3a**) compounds (72% and 77%, respectively), while bulkier trifluoromethylated diketone **5a** showed a significant loss of activity (50% conversion).

Then, the study was also performed on substrates bearing *meta* or *ortho* substitution at the phenyl ring. Hence, diketones **6–8a** were used as substrates for RasADH. However, very low yields (8%) were observed for the corresponding enantiopure methoxylated derivatives (1*S*,4*S*)-**6b** and **8b**. A special mention is deserved for *ortho*-substituted diketone **8a**, that provided hydroxy ketone **8c** at higher extent (36%) than the corresponding diol **8b**. Regarding the *m*-chloro derivative **7a**, that showed higher enzymatic conversion (22%), two different experiments were also carried out. On the one hand, the amount of enzyme was doubled leading to the formation of (1*S*,4*S*)-**7b** in 34% conversion after 24 h. On the other hand, the temperature was raised up to 40 °C, obtaining (1*S*,4*S*)-**7b** in 43% conversion after 24 h. These results showed that the substitution pattern had an effect on the enzymatic activity, achieving the most valuable results for the *para*-substituted diketones.

Next, the bioreductions of diketones **2–5a** were also attempted with the commercially available ADHs reported in Table 2. Unfortunately, no conversion was observed in any case, demonstrating the difficulty of these bioconversions, only achieved with RasADH.

Finally, preparative biotransformations up to 100-mg scale were carried out using *E. coli*/RasADH under the optimized reaction conditions (Figure 1), using THF (5% *v*/*v*) as cosolvent at 30 °C and glucose/GDH as NADPH regeneration system. After 24 h, the conversions obtained were between 40% and 90%, isolating the corresponding diastereo- and enantiopure diols (1*S*,4*S*)-**1–5,7b** in moderate to high yields. The highest yield was obtained for the unsubstituted diol (1*S*,4*S*)-**1b**, which was recovered in 73% yield, while the *m*-chloro compound (**7b**) led to the lowest yield (10%).

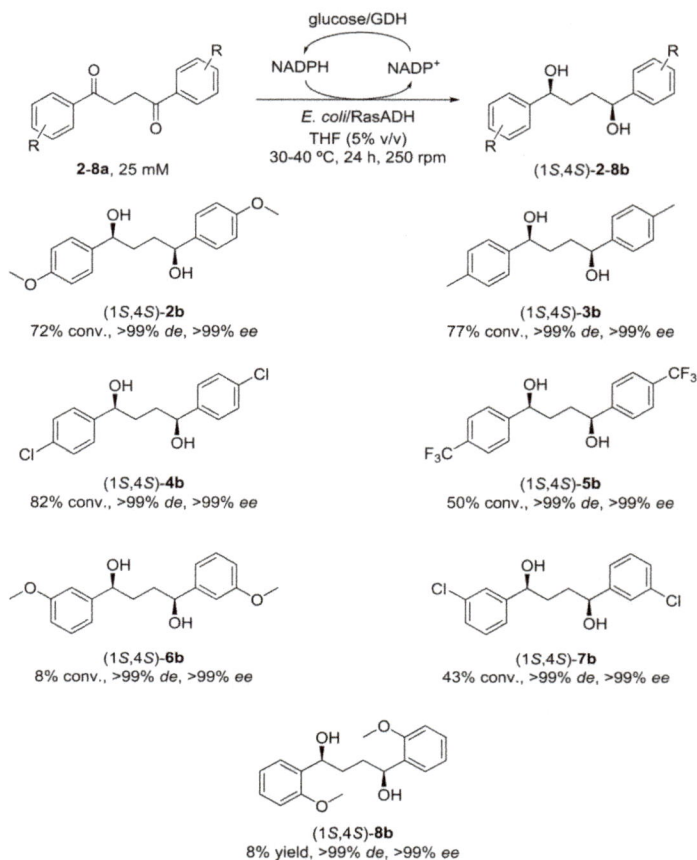

Scheme 2. Selective bioreduction of diketones 2–8a using *E. coli*/RasADH and tetrahydrofuran (THF) as cosolvent.

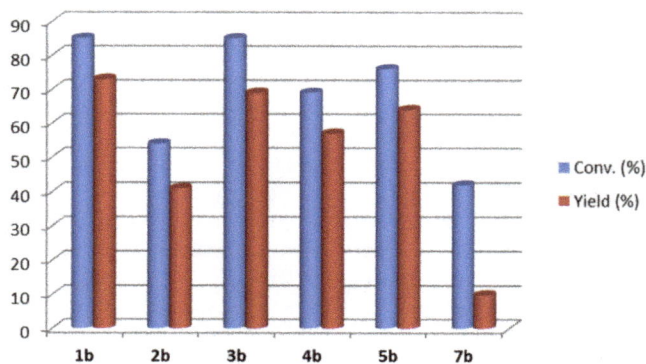

Figure 1. Conversions and isolated yields for the RasADH-catalyzed preparative biotransformations to obtain the optically active diols (1S,4S)-**1–5,7b**.

3. Conclusions

A series of 1,4-diaryl-1,4-diols were synthesized via ADH-catalyzed bioreduction from the corresponding bulky 1,4-diaryl-1,4-diketones. Among the different enzymatic preparations used, ADH from *Ralstonia* sp. (RasADH) overexpressed in *E. coli* afforded the best results in terms of conversions and diastereo- and enantiomeric excess, usually obtaining the (1*S*,4*S*)-diols, and only observing the formation of the hydroxy ketone intermediate in the case of the *o*-methoxylated derivative. The cofactor regeneration system and the addition of an organic cosolvent were studied, showing that glucose/GDH and THF were the best options. Herein we have shown that the bioreduction can be a useful alternative under mild conditions to obtain these appealing chiral compounds, which can be further employed as synthons for the preparation of other valuable derivatives.

4. Materials and Methods

As for the substrates and products studied in this contribution, HPLC separations, HPLC chromatograms of optically active 1,4-diaryl-1,4-diols and NMR spectra, please see the Supplementary Materials.

4.1. General Materials and Methods

The 4′-Substituted acetophenones and α-bromo-4′-substituted acetophenones were purchased from TCI Europe (Zwijndrecht, Belgium). NADPH as enzyme cofactor and all the other chemical reagents were obtained with the highest quality available from Sigma-Aldrich-Fluka (Steinheim, Germany). Alcohol dehydrogenases and GDH were obtained from Codexis Inc., (Redwood City, CA, USA). The production of *E. coli* overexpressed ADHs has been previously reported and these enzymatic preparations have been provided by Prof. Wolfgang Kroutil (University of Graz) [35,41,43].

NMR spectra were recorded on a Bruker AV300 MHz spectrometer (Bruker Co., Faellanden, Switzerland). All chemical shifts (δ) are given in parts per million (ppm) and referenced to the residual solvent signal as internal standard. Measurement of the optical rotation values was carried out at 590 nm on a Autopol IV Automatic polarimeter (Rudolph Research Analytical, Hackettstown, NJ, USA). High performance liquid chromatography (HPLC) analyses were performed for conversion, enantiomeric excess, and diastereomeric excess value measurements using a Hewlett Packard 1100 chromatograph UV detector at 210 nm (Agilent Technologies, Inc., Wilmington, DE, USA). As chiral columns, Chiralpak AD-H (25 cm × 4.6 mM) and Chiralpak IA (25 cm × 4.6 mM) were used (Chiral Technologies, Mainz, Germany). Thin-layer chromatography (TLC) analyses were conducted with Merck Silica Gel 60 F254 precoated plates (Merck KGaA, Darmstadt, Germany) and visualized with UV and potassium permanganate stain. Column chromatography purifications were performed using Merck Silica Gel 60 (230–400 mesh, Merck KGaA, Darmstadt, Germany).

4.2. General Procedure for the Synthesis of 1,4-Diaryl-1,4-Diketones **1–8a**

For the preparation of diketones **1–8a** we followed a similar methodology to the one described in the literature [40]. Commercial anhydrous $ZnCl_2$ (2.72 g, 20 mMol) was placed into a one-neck 250-mL round-bottom flask and dried by melting under vacuum (1 torr) at 250–350 °C for 15–20 min. After cooling under vacuum to room temperature, toluene (10 mL), diethylamine (1.03 mL, 10 mmol) and *t*BuOH (0.95 mL, 10 mmol) were successively added. The mixture was stirred until zinc chloride was fully dissolved (approx. 2 h), and then the corresponding acetophenone (8.0 mMol) and α-bromoacetophenone (5.0 mMol) were successively added. The mixture was stirred at room temperature for 4 days. After this, CH_2Cl_2 (300 mL) was added and the resulting organic phase was successively washed with an aqueous H_2SO_4 2.0 M solution (2 × 120 mL), water (1 × 150 mL) and brine (1 × 150 mL). The organic phase was dried over Na_2SO_4, filtered, and the solvents were evaporated under reduced pressure. Crude solids, except compounds **6a** and **8a** that were purified by column chromatography (Hexane/EtOAc 4:1), were purified by crystallization in absolute EtOH in order to yield the corresponding 1,4-diaryl-1,4-diketones (see Supplementary Materials) **1a** (0.69 g, 58% yield), **2a** (0.36 g, 24%), **3a** (0.33 g, 25%), **4a** (0.51 g, 33%), **5a** (0.91 g, 49%), **6a** (0.19 g, 13%), **7a** (0.79 g, 52%) and **8a** (0.11 g, 8%) [38–40,49–51].

4.3. General Procedure for the Synthesis of Racemic 1,4-Diaryl-1,4-Diols **1–8b**

Diketones **1–8a** (0.3 mMol) were dissolved in THF (2.0 mL) and NaBH$_4$ (90.8 mg, 2.4 mMol, 8.0 equiv.) was slowly added at room temperature. Reactions were stirred for 2–4 h until completion. After this, an aqueous HCl 1.0 M solution (2.0 mL) was added and the mixtures were extracted with CH$_2$Cl$_2$ (2 × 5 mL). The organic phases were dried over Na$_2$SO$_4$, filtered and the solvents were evaporated under reduced pressure. The reaction crudes were purified by column chromatography (EtOAc/hexane mixtures), isolating the racemic 1,4-diaryl-1,4-diols (±)-**1b** (61.7 mg, 85% yield), (±)-**2b** (65.2 mg, 72%), (±)-**3b** (65.6 mg, 81%), (±)-**4b** (71.9 mg, 77%), (±)-**5b** (102.1 mg, 90%), (±)-**6b** (30.0 mg, 52%), (±)-**7b** (58.3 mg, 62%) and (±)-**8b** (33.0 mg, 66%).

4.4. General Procedure for the Synthesis of Racemic 4-Hydroxy-1,4-Bis(2-methoxyphenyl)butan-1-one **8c**

Diketone **8a** (0.13 mMol) was dissolved in THF (0.9 mL) and NaBH$_4$ (21 mg, 0.54 mMol, 4.0 equiv.) was added at room temperature. The reaction was stirred for 30 min until the formation of the hydroxy ketone was observed by TLC analysis. After this, an aqueous HCl 1.0 M solution (2.0 mL) was added and the mixture was extracted with CH$_2$Cl$_2$ (2 × 5 mL). The organic layers were dried over Na$_2$SO$_4$, filtered and the solvents were evaporated under reduced pressure. The reaction crude was purified by column chromatography (EtOAc/hexane mixtures), isolating the racemic hydroxy ketone (±)-**8c** (2 mg, 5% yield).

4.5. General Procedure for the Enzymatic Conversion of 1,4-Diaryl-1,4-Diols **1–8b** Using Overexpressed E. coli/RasADH

Lyophilized *E. coli*/RasADH cells (15–30 mg), the cosolvent (2.5%–10% *v/v*), 1 mM NADP$^+$ (60 μL of a 10 mM stock solution), 50 mM glucose (60 μL of a 500 mM stock solution) and glucose dehydrogenase (5 U) were added into an Eppendorf tube containing 1,4-diaryl-1,4-diketones (**1–8a**, 25 mM) in Tris·HCl buffer 50 mM pH 7.5 (420 μL). The reaction was shaken at 30–40 °C and 250 rpm for 24–48 h. After this time, the mixture was extracted with ethyl acetate (2 × 500 μL), the organic layers separated by centrifugation (2 min, 5700× *g*), combined and finally dried over Na$_2$SO$_4$. Conversion, diastereomeric excess and enantiomeric excess values of 1,4-diaryl-1,4-diols **1–8b** were determined by HPLC (see Supplementary Materials).

4.6. General Procedure for the Bioreduction of 1,4-Diphenylbutane-1,4-Dione **1a** Using Commercial Alcohol Dehydrogenases

In a 2.0 mL Eppendorf tube, KRED (2 mg) was added to 900 μL phosphate buffer 125 mM pH 7.0 (1.25 mM mgSO$_4$, 1 mM NADP$^+$) containing 1,4-diphenylbutane-1,4-dione (**1a**, 25 mM), DMSO (25 μL) and ^{i}PrOH (100 μL). The reaction was incubated at 250 rpm and 30 °C for 24 h. Then, the mixture was extracted with ethyl acetate (2 × 500 μL), the organic layers separated by centrifugation (2 min, 5700× *g*), combined and finally dried over Na$_2$SO$_4$. Conversion, diastereomeric excess and enantiomeric excess values of 1,4-diphenylbutane-1,4-diol (**1b**) were determined by HPLC.

4.7. Preparative Bioreductions of 1,4-Diarylbutane-1,4-Diones **1–5a**, **7a**, and **8a** Using Overexpressed E. coli/RasADH

Lyophilized *E. coli*/RasADH cells (50 mg for 1,4-diketone **1a**, 20 mg for 1,4-diketones **2–4a**, 100 mg for 1,4-diketone **5a**, and 150 mg for 1,4-diketone **7a**). THF (5% *v/v*), NADP$^+$ (1 mM), glucose (50 mM), and GDH (50–100 U), were added into an Erlenmeyer flask containing a suspension of the corresponding 1,4-diketone (25 mM) in Tris·HCl buffer 50 mM pH 7.5. The reaction was incubated at 30 °C and 250 rpm for 24 h (**1a–5a**) or at 40 °C and 250 rpm for 48 h (**7a**). Then, the mixture was extracted with ethyl acetate (3 × 15 mL). The organic layers were separated by centrifugation (5 min, 4000× *g*), combined, and finally dried over Na_2SO_4. The reaction crude was purified by column chromatography (EtOAc/hexane mixtures), isolating the enantiopure (1*S*,4*S*)-diols in moderate to high yields (10%–73%).

(–)-(1*S*,4*S*)-1,4-Diphenylbutane-1,4-diol [(1*S*,4*S*)-**1b**]. Yield: 37 mg (73%). R_f = 0.36 (40% EtOAc/hexane). Mp: 74–75 °C. ^{1}H NMR (300.13 MHz, $CDCl_3$): δ 1.76–1.96 (m, 2H), 1.88–2.02 (m, 2H), 2.59 (s, 2OH), 4.72 (dd, J_{HH} = 6.5, 3.9 Hz, 2H), 7.08–7.50 (m, 10H) ppm. ^{13}C NMR (300.13 MHz, $CDCl_3$): δ 35.9 (2CH_2), 74.6 (2CH), 125.8 (4CH), 127.5 (2CH), 128.4 (4CH), 144.6 (2C) ppm. HRMS (ESI$^+$, *m/z*): calcd for $(C_{16}H_{18}NaO_2)^+$ (M + Na)$^+$ 265.1204; found 265.1199. $[\alpha]_D^{21}$ = −57.2 (*c* = 0.5, $CHCl_3$), described in the literature [52]: $[\alpha]_D^{25}$ = −59.0 (*c* = 1.0, $CHCl_3$) for *syn*-(*S*,*S*)-diol.

(–)-(1*S*,4*S*)-1,4-Bis(4-methoxyphenyl)butane-1,4-diol [(1*S*,4*S*)-**2b**]. Yield: 9 mg (41%). R_f = 0.21 (40% EtOAc/hexane). ^{1}H NMR (300.13 MHz, $CDCl_3$): δ 1.71–1.85 (m, 2H), 1.85–2.00 (m, 2H), 3.82 (s, 6H), 4.68 (m, 2H), 6.88 (d, J_{HH} = 8.5, 4H), 7.27 (d, J_{HH} = 8.1 Hz, 4H) ppm. ^{13}C NMR (300.13 MHz, $CDCl_3$): δ 34.4 (2CH_2), 53.6 (2CH_3), 81.0 (2CH), 113.9 (4CH), 127.5 (2CH), 135.2 (2C), 159.1 (2C) ppm. HRMS (ESI$^+$, *m/z*): calcd for $(C_{18}H_{22}NaO_4)^+$ (M + Na)$^+$ 325.1415; found 325.1410. $[\alpha]_D^{21}$ = −43.0 (*c* = 0.1, $CHCl_3$), described in the literature [11]: $[\alpha]_D^{22}$ = +41.6 (*c* = 1.0, $CHCl_3$) for *syn*-(*R*,*R*)-diol.

(–)-(1*S*,4*S*)-1,4-Bis(4-methylphenyl)butane-1,4-diol [(1*S*,4*S*)-**3b**]. Yield: 14 mg (69%). R_f = 0.27 (40% EtOAc/hexane). Mp: 114–115 °C. ^{1}H NMR (300.13 MHz, $CDCl_3$): δ 1.72–1.80 (m, 2H), 1.88–2.04 (m, 2H), 2.28 (s, 2OH), 2.36 (s, 6H), 4.70 (dd, J_{HH} = 6.8, 4.4 Hz, 2H), 7.16 (d, J_{HH} = 8.0 Hz, 4H), 7.24 (d, J_{HH} = 8.0 Hz, 4H) ppm. ^{13}C NMR (300.13 MHz, $CDCl_3$): δ 21.1 (2CH_3), 35.8 (2CH_2), 74.5 (2CH), 125.8 (4CH), 129.1 (4CH), 137.2 (2C), 141.7 (2C) ppm. HRMS (ESI$^+$, *m/z*): calcd for $(C_{18}H_{22}NaO_2)^+$ (M + Na)$^+$ 293.1517; found 293.1512. $[\alpha]_D^{21}$ = −45.2 (*c* = 0.2, CH_2Cl_2), described in the literature [4]: $[\alpha]_D^{25}$ = −47.0 (*c* = 1.0, $CHCl_3$) for *syn*-(*S*,*S*)-diol.

(–)-(1*S*,4*S*)-1,4-Bis(4-chlorophenyl)butane-1,4-diol [(1*S*,4*S*)-**4b**]. Yield: 11 mg (57%). R_f = 0.45 (40% EtOAc/hexane). Mp: 112–113 °C. ^{1}H NMR (300.13 MHz, $CDCl_3$): δ 1.69–2.01 (m, 4H), 2.31 (s, 2OH), 4.71 (d, J_{HH} = 5.9 Hz, 2H), 7.09–7.46 (m, 8H) ppm. ^{13}C NMR (300.13 MHz, $CDCl_3$): δ 35.8 (2CH_2), 73.8 (2CH), 127.1 (4CH), 128.5 (4CH), 133.2 (2C), 142.9 (2C) ppm. HRMS (ESI$^+$, *m/z*): calcd for $(C_{16}H_{16}Cl_2NaO_2)^+$ (M + Na)$^+$ 333.0425; found 333.0419. $[\alpha]_D^{21}$ = −22.1 (*c* = 0.2, $CHCl_3$), described in the literature [11]: $[\alpha]_D^{21}$ = +24.4 (*c* = 1.05, $CHCl_3$) for *syn*-(*R*,*R*)-diol.

(–)-(1*S*,4*S*)-1,4-Bis[4-(trifluoromethyl)phenyl]butane-1,4-diol [(1*S*,4*S*)-**5b**]. Yield: 63 mg (64%). R_f = 0.51 (40% EtOAc/hexane). Mp: 159–160 °C. ^{1}H NMR (300.13 MHz, MeOD): δ 1.67–1.77 (m, 2H), 1.78–1.92 (m, 2H), 3.30 (s, 2OH), 4.73 (apparent t, J_{HH} = 5.6 Hz, 2H), 7.49 (d, J_{HH} = 8.2 Hz, 4H), 7.60 (d, J_{HH} = 8.2 Hz, 4H) ppm. ^{13}C NMR (300.13 MHz, MeOD): δ 34.8 (2CH_2), 72.5 (2CH), 124.4 (q, J_{CF} = 271.2 Hz, 2CF_3), 124.7 (q, J_{CF} = 3.6 Hz, 4CH), 126.1 (4CH), 128.9 (q, J_{CF} = 32.0 Hz, 2C), 149.6 (2C) ppm. HRMS (ESI$^+$, *m/z*): calcd for $(C_{18}H_{16}F_6NaO_2)^+$ (M + Na)$^+$ 401.0952; found 401.0951. $[\alpha]_D^{25}$ = −20.1 (*c* = 0.4, $CHCl_3$), described in the literature [3]: $[\alpha]_D^{24}$ = +19.0 (*c* = 0.1, $CHCl_3$) for *syn*-(*R*,*R*)-diol.

(–)-(1*S*,4*S*)-1,4-Bis(3-chlorophenyl)butane-1,4-diol [(1*S*,4*S*)-**7b**]. Yield: 8 mg (10%). R_f = 0.35 (40% EtOAc/hexane). ^{1}H NMR (300.13 MHz, $CDCl_3$): δ 1.80–1.93 (m, 4H), 2.54 (s, 2OH), 4.71 (m, 2H), 7.02–7.51 (m, 8H) ppm. ^{13}C NMR (300.13 MHz, $CDCl_3$): δ 35.8 (2CH_2), 74.0 (2CH), 124.0 (2CH), 126.1 (2CH), 127.8 (2CH), 129.9 (2CH), 134.5 (2C), 146.7 (2C) ppm. HRMS (ESI$^+$, *m/z*): calcd for $(C_{16}H_{16}Cl_2NaO_2)^+$ (M + Na)$^+$ 333.0425; found 333.0426. $[\alpha]_D^{19}$ = −41.0 (*c* = 0.1, $CHCl_3$).

To obtain the hydroxy ketone **8c**, lyophilized *E. coli*/RasADH cells (80 mg), THF (5% *v/v*), NADP$^+$ (1 mM), glucose (50 mM), and GDH (50 U), were added into an Erlenmeyer flask containing a suspension of the 1,4-diketone **8a** (20 mg, 25 mM) in Tris·HCl buffer 50 mM pH 7.5. The reaction was incubated at 30 °C and 250 rpm for 24 h. Then, the mixture was extracted with ethyl acetate (3 × 15 mL) and the organic layers were separated by centrifugation (5 min, 4000× *g*), combined and finally dried over Na$_2$SO$_4$. The reaction crude was purified by column chromatography (EtOAc/hexane mixtures), isolating **8c** (2 mg, 10% yield).

4-Hydroxy-1,4-bis(2-methoxyphenyl)butan-1-one (**8c**). R_f = 0.41 (40% EtOAc/hexane). ^{1}H NMR (300.13 MHz, CDCl$_3$): δ 2.13–2.36 (m, 2H), 3.00 (s, OH), 3.05–3.29 (m, 2H), 3.85 (s, OCH$_3$), 3.88 (s, OCH$_3$), 6.79–7.07 (m, 4H), 7.25 (m, 1H), 7.38 (d, J_{HH} = 7.5 Hz, 1H), 7.40–7.51 (t, J_{HH} = 7.8 Hz, 1H), 7.69 (dd, J_{HH} = 7.7, 1.8 Hz, 1H) ppm. HRMS (ESI$^+$, *m/z*): calcd for (C$_{18}$H$_{20}$NaO$_4$)$^+$ (M + Na)$^+$ 323.1259; found 323.1260.

Supplementary Materials: The following are available online at http://www.mdpi.com/2073-4344/8/4/150/s1. Substrates and products studied in this contribution, 2. HPLC separations, 3. HPLC chromatograms of optically active 1,4-diaryl-1,4-diols, 4. NMR spectra.

Acknowledgments: Financial support from the Spanish ministry of Economy and Competitiveness (MINECO, Project CTQ2016-75752-R) is gratefully acknowledged. G.d.G. (Ramón y Cajal Program) and Á.M.-I. (FPI predoctoral fellowship) thank minECO for personal funding.

Author Contributions: Gonzalo de Gonzalo, Iván Lavandera and Vicente Gotor-Fernández conceived and designed the experiments; Gonzalo de Gonzalo and Ángela Mourelle-Insua performed the experiments; Ángela Mourelle-Insua carried out analytical measurements and analyzed the data; Gonzalo de Gonzalo, Iván Lavandera and Vicente Gotor-Fernández wrote the paper.

Conflicts of Interest: The authors declare no conflict of interest.

References

1. Robinson, A.; Aggarwal, V. Asymmetric total synthesis of solandelactone E: Stereocontrolled synthesis of the 2-ene-1,4-diol core through a lithiation–borylation–allylation sequence. *Angew. Chem.* **2010**, *122*, 6823–6825. [CrossRef]

2. Seebach, D.; Beck, A.K.; Heckel, A. TADDOLs, Their Derivatives, and TADDOL Analogues: Versatile Chiral Auxiliaries. *Angew. Chem. Int. Ed.* **2001**, *40*, 92–138. [CrossRef]

3. Denmark, S.E.; Chang, W.-T.T.; Houk, K.N.; Liu, P. Development of chiral bis-hydrazone ligands for the enantioselective cross-coupling reactions of aryldimethylsilanolates. *J. Org. Chem.* **2015**, *80*, 313–366. [CrossRef] [PubMed]

4. Chen, H.; Sweet, J.A.; Lam, K.-C.; Rheingold, A.; McGrath, D.V. Chiral amine–imine ligands based on *trans*-2,5-disubstituted pyrrolidines and their application in the palladium-catalyzed allylic alkylation. *Tetrahedron Asymmetry* **2009**, *20*, 1672–1682. [CrossRef]

5. Melchiorre, P.; Jorgensen, K.A. Direct enantioselective Michael addition of aldehydes to vinyl ketones catalyzed by chiral amines. *J. Org. Chem.* **2003**, *68*, 4151–4157. [CrossRef] [PubMed]

6. Foubelo, F.; Nájera, C.; Yus, M. Catalytic asymmetric transfer hydrogenation of ketones: Recent advances. *Tetrahedron Asymmetry* **2015**, *26*, 769–790. [CrossRef]

7. Hynes, J.T.; Klinman, J.P.; Limbach, H.-H.; Schowen, R.L. (Eds.) *Hydrogen Transfer Reactions*; Wiley-VCH: Weinheim, Germany, 2007; Volumes 1–4.

8. Zheng, L.-S.; Llopis, Q.; Echeverria, P.-G.; Férand, C.; Guillamot, G.; Phansavath, P.; Ratovelomanana-Vidal, V. Asymmetric transfer hydrogenation of (hetero)arylketones with tethered Rh(III)−N-(*p*-tolylsulfonyl)-1,2-diphenylethylene-1,2-diamine complexes: Scope and limitations. *J. Org. Chem.* **2017**, *82*, 5607–5615. [CrossRef] [PubMed]

9. Corey, E.J.; Helal, C.J. Reduction of carbonyl compounds with chiral oxazaborolidine catalyst: A new paradigm for enantioselective catalysis and a powerful new synthetic method. *Angew. Chem. Int. Ed.* **1998**, *37*, 1986–2012. [CrossRef]

10. Mathre, D.J.; Thomson, A.S.; Douglas, A.W.; Hoogsteen, K.; Carroll, J.D.; Corley, E.G.; Grabowski, E.J.J. A practical process for the preparation of tetrahydro-1-methyl-3,3-diphenyl-1*H*,3*H*-pyrrolo[1,2-c][1,3,2]oxazaborole-borane. A highly enantioselective stoichiometric and catalytic reducing agent. *J. Org. Chem.* **1993**, *58*, 2880–2888. [CrossRef]

11. Li, X.; Zhao, G.; Cao, W.-G. An efficient method for catalytic asymmetric reduction of diketones and application of synthesis to chiral 2,5-diphenylpyrrolidine and 2,5-diphenylthiolane. *Chin. J. Chem.* **2006**, *24*, 1402–1405. [CrossRef]

12. Aldous, D.J.; Dutton, W.M.; Steel, P.G. A simple enantioselective preparation of (2*S*,5*S*)-2,5-diphenylpyrrolidine and related diaryl amines. *Tetrahedron Asymmetry* **2000**, *11*, 2455–2462. [CrossRef]

13. Hudlicky, T.; Reed, J.W. Applications of biotransformations and biocatalysis to complexity generation in organic synthesis. *Chem. Soc. Rev.* **2009**, *38*, 3117–3132. [CrossRef] [PubMed]

14. Clouthier, C.M.; Pelletier, J.M. Expanding the organic toolbox: A guide to integrating biocatalysis in synthesis. *Chem. Soc. Rev.* **2012**, *41*, 1585–1605. [CrossRef] [PubMed]

15. Milner, S.E.; Maguire, A.R. Recent trends in whole cell and isolated enzymes in enantioselective synthesis. *Arkivoc* **2012**, *2012*, 321–382.

16. Torrelo, G.; Hanefeld, U.; Hollmann, F. Biocatalysis. *Catal. Lett.* **2015**, *145*, 309–345. [CrossRef]

17. Albarrán-Velo, J.; González-Martínez, D.; Gotor-Fernández, V. Stereoselective Biocatalysis. A mature technology for the asymmetric synthesis of pharmaceutical building blocks. *Biocatal. Biotransform.* **2018**, *36*, 102–130. [CrossRef]

18. Mattson, A.; Öhrner, N.; Hult, K.; Norin, T. Resolution of diols with C_2-symmetry by lipase catalysed transesterification. *Tetrahedron Asymmetry* **1993**, *4*, 925–930. [CrossRef]

19. Nagai, H.; Morimoto, T.; Achiwa, K. Facile enzymatic synthesis of optically active 2,5-hexanediol derivatives and its application to the preparation of optically pure cyclic sulfate for chiral ligands. *Synlett* **1994**, *1994*, 289–290. [CrossRef]

20. Caron, G.; Kazlauskas, R.J. Isolation of racemic 2,4-pentanediol and 2,5-hexanediol from commercial mixtures of racemic and *meso* isomers by way of cyclic sulfites. *Tetrahedron Asymmetry* **1994**, *6*, 657–664. [CrossRef]

21. Persson, B.A.; Larsson, A.L.E.; Le Ray, M.; Bäckvall, J.-E. Ruthenium- and enzyme-catalyzed dynamic kinetic resolution of secondary alcohols. *J. Am. Chem. Soc.* **1999**, *121*, 1645–1650. [CrossRef]

22. Persson, B.A.; Huerta, F.; Bäckvall, J.-E. Dynamic kinetic resolution of secondary diols via coupled ruthenium and enzyme catalysis. *J. Org. Chem.* **1999**, *64*, 5237–5240. [CrossRef]

23. Edin, M.; Bäckvall, J.-E. On the mechanism of the unexpected facile formation of *meso*-diacetate products in enzymatic acetylation of alkanediols. *J. Org. Chem.* **2003**, *68*, 2216–2222. [CrossRef] [PubMed]

24. Martín-Matute, B.; Edin, M.; Bäckvall, J.-E. Highly efficient synthesis of enantiopure diacetylated C_2-symmetric diols by ruthenium- and enzyme-catalyzed dynamic kinetic asymmetric transformation (DYKAT). *Chem. Eur. J.* **2006**, *12*, 6053–6061. [CrossRef] [PubMed]

25. Borén, L.; Leijondahl, K.; Bäckvall, J.-E. Dynamic kinetic asymmetric transformation of 1,4-diols and the preparation of trans-2,5-disubstituted pyrrolidines. *Tetrahedron Lett.* **2009**, *50*, 3237–3240. [CrossRef]

26. Martín-Matute, B.; Bäckvall, J.-E. Ruthenium- and enzyme-catalyzed dynamic kinetic asymmetric transformation of 1,4-diols: Synthesis of γ-hydroxy ketones. *J. Org. Chem.* **2004**, *69*, 9191–9195. [CrossRef] [PubMed]

27. Yang, B.; Lihammar, R.; Bäckvall, J.-E. Investigation of the impact of water on the enantioselectivity displayed by CALB in the kinetic resolution of δ-functionalized alkan-2-ol derivatives. *Chem. Eur. J.* **2014**, *20*, 13517–13521. [CrossRef] [PubMed]

28. Hollmann, F.; Arends, I.W.C.E.; Holtmann, D. Enzymatic reductions for the chemist. *Green Chem.* **2011**, *13*, 2285–2313. [CrossRef]

29. Magano, J.; Dunetz, J.R. Large-scale carbonyl reductions in the pharmaceutical industry. *Org. Process Res. Dev.* **2012**, *16*, 1156–1184. [CrossRef]

30. Nealon, C.M.; Musa, M.M.; Patel, J.M.; Phillips, R.S. Controlling substrate specificity and stereospecificity of alcohol dehydrogenases. *ACS Catal.* **2015**, *5*, 2100–2114. [CrossRef]

31. Kratzer, R.; Woodley, J.M.; Nidetzky, B. Rules for biocatalyst and reaction engineering to implement effective, NAD(P)H-dependent, whole cell bioreductions. *Biotechnol. Adv.* **2015**, *33*, 1641–1655. [CrossRef] [PubMed]

32. Kurina-Sanz, M.; Bisogno, F.R.; Lavandera, I.; Orden, A.A.; Gotor, V. Promiscuous substrate-binding explains the enzymatic stereo- and regiocontrolled synthesis of enantiopure hydroxy ketones and diols. *Adv. Synth. Catal.* **2009**, *351*, 1842–1848. [CrossRef]

33. Chen, Y.; Chen, C.; Wu, X. Dicarbonyl reduction by single enzyme for the preparation of chiral diols. *Chem. Soc. Rev.* **2012**, *41*, 1742–1753. [CrossRef] [PubMed]

34. Goldberg, K.; Edegger, K.; Kroutil, W.; Liese, A. Overcoming the thermodynamic limitation in asymmetric hydrogen transfer reactions catalyzed by whole cells. *Biotechnol. Bioeng.* **2006**, *95*, 192–198. [CrossRef] [PubMed]

35. De Gonzalo, G.; Lavandera, I.; Faber, K.; Kroutil, W. Enzymatic reduction of ketones in "micro-aqueous" media catalyzed by ADH-A from *Rhodococcus ruber*. *Org. Lett.* **2007**, *9*, 2163–2166. [CrossRef] [PubMed]

36. Machielsen, R.; Leferink, N.G.H.; Hendriks, A.; Brouns, S.J.J.; Hennemann, H.-G.; Daussmann, T.; Oost, J. Laboratory evolution of *Pyrococcus furiosus* alcohol dehydrogenase to improve the production of (2S,5S)-hexanediol at moderate temperatures. *Extremophiles* **2008**, *12*, 587–594. [CrossRef] [PubMed]

37. Müller, M.; Katzberg, M.; Bertau, M.; Hummel, W. Highly efficient and stereoselective biosynthesis of (2S,5S)-hexanediol with a dehydrogenase from *Saccharomyces cerevisiae*. *Org. Biomol. Chem.* **2010**, *8*, 1540–1550. [CrossRef] [PubMed]

38. Mizar, P.; Wirth, T. Flexible stereoselective functionalizations of ketones through umpolung with hypervalent iodine reagents. *Angew. Chem. Int. Ed.* **2014**, *53*, 5993–5997. [CrossRef] [PubMed]

39. Hua, G.; Henry, J.B.; Li, Y.; Mount, A.R.; Slawin, A.M.Z.; Woollins, J.D. Synthesis of novel 2,5-diarylselenophenes from selenation of 1,4-diarylbutane-1,4-diones or methanol/arylacetylenes. *Org. Biomol. Chem.* **2010**, *8*, 1655–1660. [CrossRef] [PubMed]

40. Nevar, N.M.; Kel'in, A.V.; Kulinkovich, O.G. One step preparation of 1,4-diketones from methyl ketones and α-bromomethyl ketones in the presence of $ZnCl_2$ • *t*-BuOH• Et_2NR as a condensation agent. *Synthesis* **2000**, *9*, 1259–1262. [CrossRef]

41. Lavandera, I.; Kern, A.; Ferreira-Silva, B.; Glieder, A.; de Wildeman, S.; Kroutil, W. Stereoselective bioreduction of bulky-bulky ketones by a novel ADH from *Ralstonia* sp. *J. Org. Chem.* **2008**, *73*, 6003–6005. [CrossRef] [PubMed]

42. Leuchs, S.; Greiner, L. Alcohol dehydrogenase from *Lactobacillus brevis*: A versatile robust catalyst for enantioselective transformations. *Chem. Biochem. Eng. Q.* **2011**, *25*, 267–281.

43. Lavandera, I.; Kern, A.; Resch, V.; Ferreira-Silva, B.; Glieder, A.; Fabian, W.M.F.; de Wildeman, S.; Kroutil, W. One-way biohydrogen transfer for oxidation of *sec*-alcohols. *Org. Lett.* **2008**, *10*, 2155–2158. [CrossRef] [PubMed]

44. Heiss, C.; Laivenieks, M.; Zeikus, J.G.; Phillips, R.S. Mutation of cysteine-295 to alanine in secondary alcohol dehydrogenase from *Thermoanaerobacter ethanolicus* affects the enantioselectivity and substrate specificity of ketone reductions. *Bioorg. Med. Chem.* **2001**, *9*, 1659–1666. [CrossRef]

45. Stampfer, W.; Kosjek, B.; Moitzi, C.; Kroutil, W.; Faber, K. Biocatalytic asymmetric hydrogen transfer. *Angew. Chem. Int. Ed.* **2002**, *41*, 1014–1017. [CrossRef]

46. Abu, R.; Woodley, J. Application of enzyme coupling reactions to shift thermodynamically limited biocatalytic reactions. *ChemCatChem* **2015**, *7*, 3094–3105. [CrossRef]

47. Man, H.; Kędziora, K.; Kulig, J.; Frank, A.; Lavandera, I.; Gotor-Fernández, V.; Rother, D.; Hart, S.; Turkenburg, J.P.; Grogan, G. Structures of alcohol dehydrogenases from *Ralstonia* and *Sphingobium* spp. reveal the molecular basis for their recognition of 'bulky–bulky' ketones. *Top. Catal.* **2014**, *57*, 356–365. [CrossRef]

48. García-Cerrada, S.; Redondo-Gallego, L.; Martínez-Olid, F.; Rincón, J.A.; García-Losada, P. Practical manufacture of 4-alkyl-4-aminocyclohexylalcohols using ketoreductases. *Org. Process Res. Dev.* **2017**, *21*, 779–784. [CrossRef]

49. Ceylan, M.; Gürdere, M.B.; Budak, Y.; Kazaz, C.; Seçen, H. One-step preparation of symmetrical 1,4-diketones from α-halo ketones in the presence of $Zn-I_2$ as a condensation agent. *Synthesis* **2004**, *35*, 1750–1754. [CrossRef]

50. Fujita, K.; Shinokubo, H.; Oshima, K. Highly diastereoselective tandem reduction–allylation reactions of 1,4-diketones with zirconocene–olefin complexes. *Angew. Chem. Int. Ed.* **2003**, *42*, 2550–2552. [CrossRef] [PubMed]

51. Xuan, J.; Feng, Z.-J.; Chen, J.-R.; Lu, L.-Q.; Xiao, W.-J. Visible-light-induced C-S bond activation: Facile access to 1,4-diketones from β-ketosulfones. *Chem. Eur. J.* **2014**, *20*, 3045–3049. [CrossRef] [PubMed]

52. Domin, D.; Benito-Garagorri, D.; Mereiter, K.; Hametner, C.; Fröhlich, J.; Kirchner, K. Synthesis and characterization of new chiral palladium β-diimine complexes. *J. Organomet. Chem.* **2007**, *692*, 1048–1057. [CrossRef]

catalysts

MDPI

Article

Use of *Lactobacillus rhamnosus* (ATCC 53103) as Whole-Cell Biocatalyst for the Regio- and Stereoselective Hydration of Oleic, Linoleic, and Linolenic Acid

Stefano Serra * and Davide De Simeis

Consiglio Nazionale delle Ricerche (C.N.R.), Istituto di Chimica del Riconoscimento Molecolare,
Via Mancinelli 7, 20131 Milano, Italy; dav.biotec01@gmail.com
* Correspondence: stefano.serra@cnr.it or stefano.serra@polimi.it; Tel.: +39-02-2399-3076

Received: 22 February 2018; Accepted: 8 March 2018; Published: 9 March 2018

Abstract: Natural hydroxy fatty acids are relevant starting materials for the production of a number of industrial fine chemicals, such as different high-value flavour ingredients. Only a few of the latter hydroxy acid derivatives are available on a large scale. Therefore, their preparation by microbial hydration of unsaturated fatty acids, affordable from vegetable oils, is a new biotechnological challenge. In this study, we describe the use of the probiotic bacterium *Lactobacillus rhamnosus* (ATCC 53103) as whole-cell biocatalyst for the hydration of the most common unsaturated octadecanoic acids, namely oleic acid, linoleic acid, and linolenic acid. We discovered that the addition of the latter fatty acids to an anaerobic colture of the latter strain, during the early stage of its exponential growth, allows the production of the corresponding mono-hydroxy derivatives. In these experimental conditions, the hydration reaction proceeds with high regio- and stereoselectivity. Only 10-hydroxy derivatives were formed and the resulting (*R*)-10-hydroxystearic acid, (*S*)-(12Z)-10-hydroxy-octadecenoic acid, and (*S*)-(12Z,15Z)-10-hydroxy-octadecadienoic acid were obtained in very high enantiomeric purity (ee > 95%). Although overall conversions usually do not exceed 50% yield, our biotransformation protocol is stereoselective, scalable, and holds preparative significance.

Keywords: hydratase; oleic acid; linoleic acid; linolenic acid; hydroxy fatty acids; stereoselective biotransformation; *Lactobacillus rhamnosus*

1. Introduction

Hydroxy fatty acids (HFAs) are important chemicals widely used for a number of applications, such as starting materials for biodegradable polymers, lubricants, emulsifiers, drugs, cosmetic ingredients, and flavours [1–4]. A very large number of HFAs have been identified in nature, but only a few of them are available in industrially significant amounts. This is the case of ricinoleic acid **1** (12-hydroxy-9-*cis*-octadecenoic acid) that is commonly used in industry as it is the major fatty acid component of castor oil (Figure 1). Consequently, the supply of other HFAs is usually achieved by hydration of the unsaturated fatty acids (UFAs), straightforwardly available from natural sources. A large number of different (UFAs), possessing multiple double bonds are components of vegetable oils or fish fats. Therefore, the preparation of many HFAs is possible. Unfortunately, even if this kind of reaction can be efficiently performed by a number of chemical means, the latter processes are usually performed using harsh experimental conditions (strong acid catalysts, high temperatures) that lack of stereochemical control. Thus, complex mixtures of isomers are usually formed. In addition, according to the European and US legislation the obtained HFAs are considered as artificial and are no longer exploitable as starting precursors for the preparation of natural flavours [4].

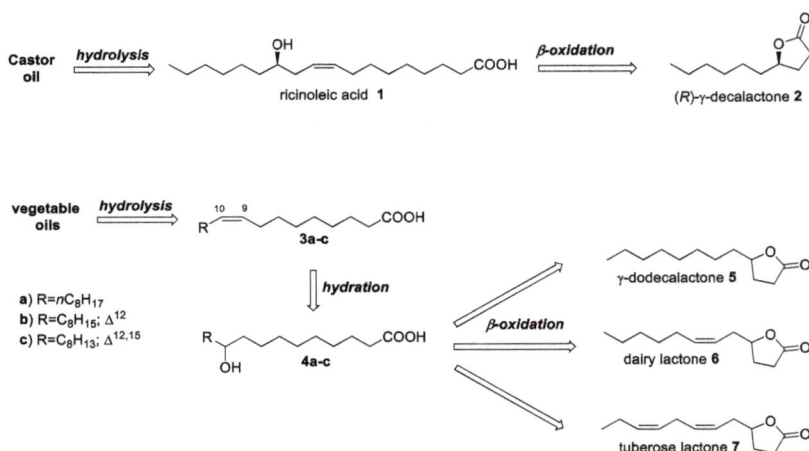

Figure 1. Synthesis of natural (+)-(R)-gamma-decalactone from castor oil and the prospective synthesis of natural C_{12} lactones gamma-dodecalactone, dairy lactone and tuberose lactone from vegetable oils through exploitation of UFA hydration reactions.

In this context, the most relevant application involving HFAs is their microbial degradation to lactones. Usually, this process is conveniently performed by means of different yeast strains [5,6] that use these fatty acids as a carbon source and transform them through many cycles of β-oxidation in the corresponding gamma or delta lactones. The majority of the fatty acid-deriving lactones (C_9–C_{12}) are of high interest in F&F industry because are widely used for food flavouring. These compounds are not available by extraction from natural sources, therefore, the only affordable way for their preparation, in natural form, is the biotransformation of natural precursors, such as natural HFAs. Lactones obtained by this way can be labelled as natural and, thus, possess much higher commercial value, with prices ranging from 300 to 3000 €/Kg.

A reliable process based on the microbial transformation of castor oil [6], secures the production of natural (+)-(R)-gamma-decalactone **2** (Figure 1). On the contrary, there are no affordable natural HFA precursors for other sought-after C_{12} gamma lactones. In principle, the most straightforward and challenging way for their synthesis is based on the enzymatic hydration of the very common Δ^{9-10} unsaturated fatty acids of type **3**, in order to produce the corresponding 10-hydroxy derivatives **4**. The hydrolysis of a number of vegetable oils affords this kind of fatty acids, such as oleic acid **3a**, linoleic acid **3b**, and α-linolenic acid **3c**. Therefore, some high-value gamma lactones [7], for example gamma dodecalactone **5**, the structurally-related dodecelactone **6** (dairy lactone), and dodecadienelactone **7** [8] (tuberose lactone) could be prepared following this approach.

The hydration reaction of unsaturated fatty acids was discovered in the early 1960s, during a study on the hydration of oleic acid using a *Pseudomonas* strain [9,10]. Afterwards, a number of other microorganisms proved to be able to perform this transformation [11–25] but the enzymes responsible for the hydration step (oleate hydratases) have been characterized only recently [26], receiving growing attention both from chemists and biologists [27–37]. It is worth noting that different putative oleate hydratase have been cloned from a number of bacteria strains, but none of them have been used for the industrial synthesis of HFAs. To date, the transformation of natural UFAs in HFAs, at the preparative scale, has been achieved only by means of whole-cell based procedures. These studies take advantage of the high hydratase activity of some specific bacteria, regardless of the biosafety level they belong. Since we are interested in developing a reliable process for the synthesis of HFAs, to be employed as starting materials for the production of natural food flavours, we limited our study to microorganisms belonging to biosafety level 1 and recognized with technological beneficial use in foods [38].

As different studies have reported the 10-hydratase activity of some *Lactobacillus* species [2,21,22], namely *acidophilus*, *plantarum*, *casei*, *paracasei*, *lactis*, *delbrueckii*, *reuteri*, *bulgaricus*, and *rhamnosus* LGG, we selected *Lactobacillus rhamnosus* LGG (ATCC 53103) as the most suitable whole-cell biocatalyst for the above mentioned hydration reaction. Actually, this microorganism has been isolated from the intestinal tract of healthy human beings and is available on the market in lyophilized form since is currently used as a probiotic [39] and has been already employed for whole-cell biotransformation processes [40]. Being regarded as beneficial for human health, the use of the latter strain does not involve any safety concerns and can be employed in industrial processes for food flavour production.

In the present work, we describe the use of this microorganism as a whole-cell biocatalyst for the hydration of the most common unsaturated octadecanoic acids, namely oleic acid, linoleic acid, and linolenic acid. More specifically, we study a preparative procedure for their conversion in the corresponding 10-hydroxy-derivatives. Our studies are also finalized to determine the regio and stereoselectivity of the hydration step. As linoleic and linolenic acids possess two and three double bonds, respectively, the biotransformation can affect up to three position of the fatty acids. Even if only the mono-hydroxy derivatives are formed, the reaction can afford different regioisomers, each ones as *R* or *S* enantiomers. This part of the study was performed by GC-MS analysis of the biotransformation mixtures and by NMR analysis of specific derivatives of the isolated hydroxy acids. The results showed that the investigated reaction is completely region- and stereoselective affording (*R*)-10-hydroxystearic acid, (*S*)-(12Z)-10-hydroxy-octadecenoic acid, and (*S*)-(12Z,15Z)-10-hydroxy-octadecadienoic acid as sole products.

2. Results and Discussion

According to the patent literature that described the isolation and the culture conditions of *Lactobacillus rhamnosus* ATCC 53103 [39], we have grown the latter strain in anaerobic flasks, at 37 °C and using MSR broth as a medium. Preliminary biotransformation experiments were performed by adding the suitable fatty acid (3 g/L), dissolved in ethanol (<0.5% final concentration), to an active culture of the *Lactobacillus*. In order to exclude growth variability due to quorum-sensing effect [41], we used the same bacterial inoculum for each trials (7.5×10^7 CFU/mL). The fatty acids were added at once, at different stages of the culture growth and the formation of the corresponding HFAs was detected by TLC analysis. We observed that each one of the three fatty acids markedly inhibited the microbial growth to such an extent that the addition of the UFAs within the first hour after the inoculum allowed obtaining neither a proper bacterial culture nor the wanted HFAs derivatives. Otherwise, when the microorganism is in the stationary phase, the addition of the UFAs produces a minor amount of the corresponding HFAs derivatives.

These observations agree with Hagen's work [26], in which the expression of the oleate hydratase from *Elizabethkingia meningoseptica* was induced by the presence of oleic acid. As a consequence, the hydration reaction can be properly achieved adding the fatty acid during the exponential phase of the bacteria growth. In order to establish the best biotransformation conditions, we describe the *Lactobacillus rhamnosus* ATCC 53103 growth curve by sampling a flask culture, prepared as described above and measuring its optical density (OD_{600}) at regular time interval (Figure 2).

Combining the latter data with the preliminary results of the flask-based biotransformation experiments, we selected as the most suitable moment for the fatty acids addition the first part of the exponential growth phase (3 h for the flask experiments). Since the microorganism under study produces lactic acid by glucose catabolism, we took advantage of the deriving pH variation to define the exponential phase span. In accord with the main aim of our work, we exploited the data described above in order to scale up the biotransformation.

The process was performed in fermenter, each experiment on a scale superior to one litre and using a fatty acid concentration of 5 g/L. The pH was controlled by dropwise addition of an aqueous solution of either acetic acid or ammonia. The investigated *Lactobacillus* strain produce a defined amount of lactic acid, directly proportional to the glucose content of the medium.

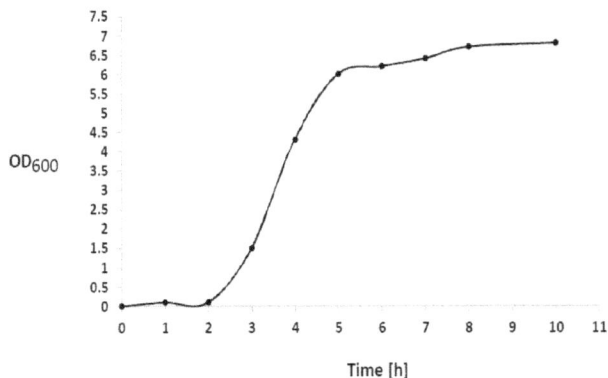

Figure 2. Growth curve of *Lactobacillus rhamnosus* ATCC 53103. Conditions: anaerobic flask, MRS broth, 37 °C, inoculum 7.5×10^7 CFU/mL, 130 rpm.

In the experimental conditions described, the ammonia uptake necessary to keep the pH at the fixed value of 6.2 was 11 mmol per gram of glucose. About 2–6 h after the inoculum of the microbial precolture, the fermentation entered its exponential phase of growth as indicated by the start of the automatic addition of the base. After consumption of about one fourth of the initial glucose content, the fatty acid was added at once. In order to allow the maximum hydratase production, the microbial growth was then forced by addition of further glucose as soon as that contained in the medium ran out.

After 48 h since the fatty acid addition, the TLC analysis showed that the hydration reaction did not proceed further. The biotransformation was stopped and both the unreacted UFAs and the HFAs formed were isolated by chromatography. The biotransformation of each one of the three selected UFAs was performed in triplicate and the results obtained are schematically described in Figure 3. The indicated yields correspond to the average of three different experimental values.

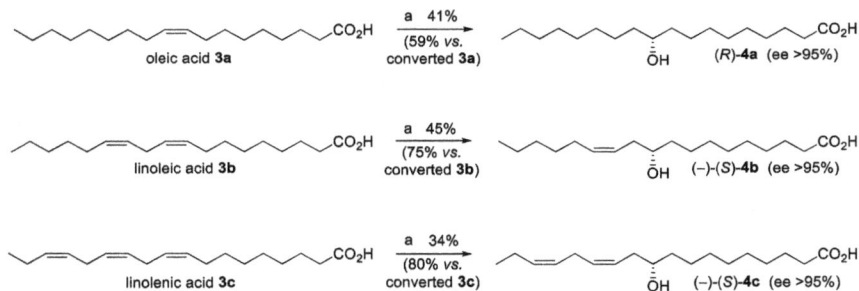

Figure 3. Whole-cell based biotransformation of oleic, linoleic and linolenic acid using *Lactobacillus rhamnosus* (ATCC 53103). Experimental conditions: (a) anaerobic fermentation, MRS broth, 37 °C, pH 6.2, 170 rpm, fatty acid concentration 5 g/L.

The perusal of the obtained data allows drawing some relevant conclusions. First, the number of the double bonds present on the starting fatty acid has a limited influence on the absolute yields of the obtained HFAs. This value range from 34% for linolenic acid to 45% for linoleic acid whereas oleic acid affords the corresponding HFA in 41% yield. Longer contact times or lower UFA concentrations did not increase the yields. It seems possible that the formed HFAs could act as inhibitors of the hydration reaction itself and, thus, overall yields are the result of the equilibrium of the hydration/dehydration reactions. Otherwise, the yields versus transformed UFAs indicated that the investigated microorganism does not transform the substrates in derivatives different from HFAs, with the exception of oleic acid for which we detected a minor and unspecific partial degradation.

The isolated HFAs were characterized by NMR, ESI-MS and GC-MS analysis. The results confirmed the chemical structures represented in Figure 3, indicating that *Lactobacillus rhamnosus* hydrates oleic, linoleic, and linolenic acids to give the corresponding 10-hydroxyderivatives, namely 10-hydroxystearic, (12Z)-10-hydroxy-octadecenoic, and (12Z,15Z)-10-hydroxy-octadecadienoic acids, respectively.

The recorded ^{1}H- and ^{13}C-NMR spectra are in very good agreement with those previously reported in the literature [42] for the same HFAs.

Furthermore, the three HFAs were derivatized by means of the sequential treatment with diazomethane followed by acetic anhydride in pyridine. The obtained derivatives **8a–c** (Figure 4) appear as sharp, well-resolved peaks by GC-MS analysis, whose electron impact spectrums share in common the ions showing m/z 201 and 169. The latter fragmentation patterns are most likely formed from alpha cleavage with respect to the 10-acetoxy group and, thus, their presence give a strict confirmation of the hydroxy group position on the fatty acid chain. Finally, the GC-MS analysis of each one of the latter derivatives showed the presence of a single peak, confirming that only the 10-hydroxy derivatives were formed, regardless of the number (or position) of the other double bonds. The same analytical procedure was repeated using samples of the three crude biotransformation mixtures. The results of the latter analyses cannot be effected by the purification procedures and confirmed again the exclusive presence of the above-described 10-hydroxy derivatives.

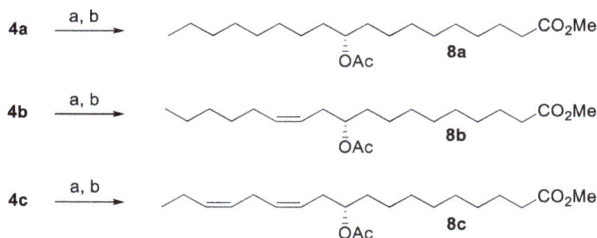

Figure 4. Transformation of 10-hydroxystearic, (12Z)-10-hydroxy-octadecenoic and (12Z,15Z)-10-hydroxy-octadecadienoic acids **4a–c** to the corresponding derivatives **8a–c**. Reagents and conditions: (a) CH$_2$N$_2$, Et$_2$O, 0 °C; (b) Ac$_2$O/Py, DMPA cat., RT.

Another important topic of our work concerns the determination of the stereoselectivity related to the hydration reaction. The isolated (12Z)-10-hydroxy-octadecenoic and (12Z,15Z)-10-hydroxy-octadecadienoic acids both showed the negative optical rotation value of −6.4 and −4.7, respectively, corresponding to (S) absolute configuration. Moreover, 10-hydroxystearic acid possesses an optical rotation value almost equal to zero and its configuration is not assignable through the latter measurement. In order to determine the missing assignment and to measure the enantiomeric purity of all three HFAs obtained by *Lactobacillus rhamnosus* biotransformation, we derivatized them according to the Rosazza procedure [43] (Figure 5).

The latter analytical method was developed for ascertaining the stereochemical purity of 10-HSA and is based on the ^{1}H-NMR analysis of the diastereoisomeric (S)-O-acetylmandelate esters of the corresponding methyl-10-hydroxystearate. The methyl ester signals due to (R) and (S)-10-hydroxystearic acid derivatives gives two well-resolved singlets at 3.66 and 3.67 ppm, respectively, whose relative peak areas indicate the corresponding isomeric ratio.

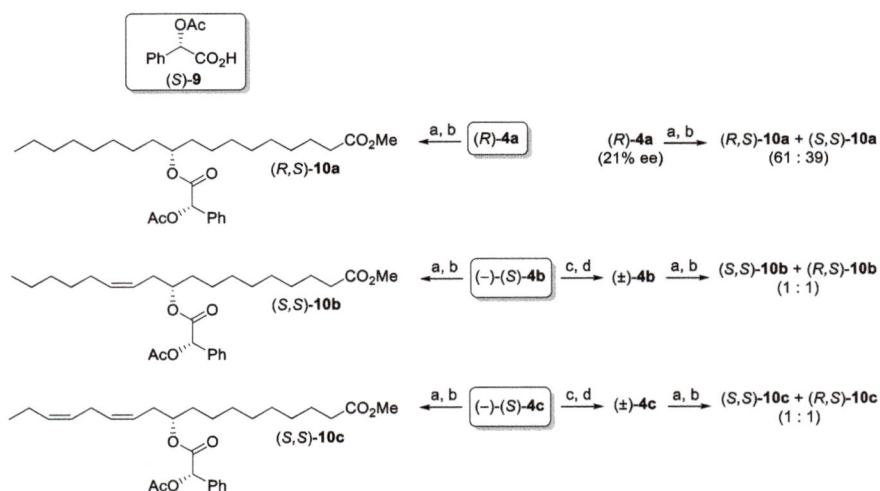

Figure 5. Synthesis of racemic (12Z)-10-hydroxy-octadecenoic and (12Z,15Z)-10-hydroxy-octadecadienoic acids and the transformation of the HFAs **4a–c** into the corresponding (S)-O-acetylmandelate esters **10a–c**. Reagents and conditions: (a) CH_2N_2, Et_2O, 0 °C; (b) (S)-**9**, DCC, cat. DMAP, CH_2Cl_2 r.t.; (c) DMSO, ClCOOCl, Et_3N, CH_2Cl_2, −70 °C; (d) $NaBH_4$, MeOH, 0 °C.

Concerning 10-hydroxystearic acid, we used as a reference standard a (R)-10-hydroxystearic acid sample having 21% ee, obtained by baker's yeast-mediated oleic acid hydration [25]. Accordingly, both the 10-HSA obtained by biotransformation and the above-mentioned (R)-standard were treated with diazomethane and then converted in the corresponding (S)-O-acetylmandelate esters **10a**. The ^{1}H-NMR analysis (Figure 6) of these two derivatives showed the presence of a 61:39 mixture of (R,S)-**10a** and (S,S)-**10a** for the reference standard (sample **a**) and a 98:2 mixture of (R,S)-**10a** and (S,S)-**10a** for the sample of 10-HSA produced by means of *Lactobacillus rhamnosus* (sample **b**). This experiment attests unambiguously that the latter microorganism hydrates oleic acid with complete regio and stereospecificity affording (R)-10-hydroxystearic acid with ee > 95%.

In order to obtain reference standard samples of racemic hydroxy acids **4b** and **4c**, we oxidized the corresponding (S)-enantiomers obtained by biotransformation. The obtained ketones were then reduced to racemic alcohols. As the intermediate ketones are β,γ-unsaturated, to avoid isomerization, the oxidation was performed at −70 °C, using Swern conditions [44]. The ketones were not purified and were immediately reduced using $NaBH_4$ in methanol.

Again, both the 10-hydroxy acids obtained by the biotransformation procedures and the above-mentioned racemic standards were treated with diazomethane and then converted in the corresponding (S)-O-acetylmandelate esters. Esters **10b** and **10c** were prepared using racemic **4b** and **4c**, respectively. The ^{1}H-NMR analysis of these two esters showed that the hydrogens linked to the carbon bearing the acetoxy group give two well-resolved singlets at 5.890 and 5.878 ppm (Figure 6, sample **c** and sample **e**). The (S)-O-acetylmandelate esters of (S)-(12Z)-10-hydroxy-octadecenoic and (S)-(12Z,15Z)-10-hydroxy-octadecadienoic acid methyl esters are responsible for the signals at 5.878 ppm. As a consequence, comparing the relative peak areas measured for the diastereoisomeric compounds (S,S)-**10b**/(R,S)-**10b** and (S,S)-**10c**/(R,S)-**10c** we determined the enantiomeric purity of (S)-(12Z)-10-hydroxy-octadecenoic and (S)-(12Z,15Z)-10-hydroxy-octadecadienoic acids.

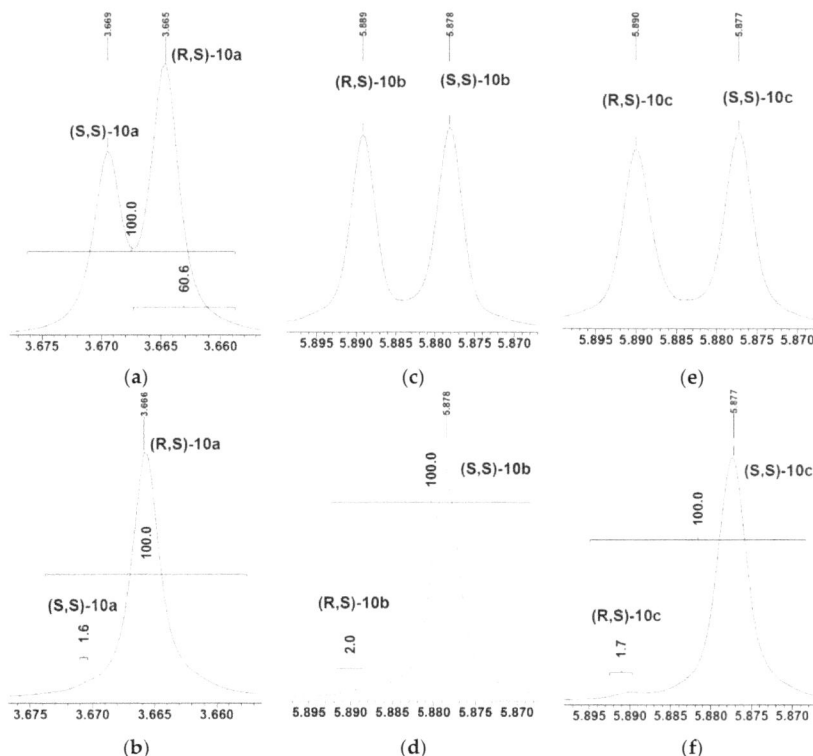

Figure 6. ^{1}H-NMR analysis of the diastereoisomeric mixtures of (*S*)-*O*-acetylmandelate esters **10a–c** deriving from 10-hydroxystearic (**4a**), (12Z)-10-hydroxy-octadecenoic (**4b**) and (12Z,15Z)-10-hydroxy-octadecadienoic (**4c**) acid samples either prepared as reference standards (samples **a**, **c**, and **e**) or obtained by *Lactobacillus rhamnosus*-mediated hydration reactions (samples **b**, **d**, **f**). Samples description: (**a**) ester **10a** prepared using (*R*)-**4a** having 21% ee; (**b**) ester **10a** prepared using **4a** obtained by hydration of oleic acid; (**c**) ester **10b** prepared using racemic **4b**; (**d**) ester **10b** obtained by the hydration of linoleic acid; (**e**) ester **10c** prepared using racemic **4c**; and (**f**) ester **10c** prepared using **4c** obtained by the hydration of linolenic acid.

Accordingly, ester **10b**, prepared using **4b** obtained by the hydration of linoleic acid (sample **d**) and ester **10c**, prepared using **4c** obtained by the hydration of linolenic acid (sample **f**) showed a diastereoisomeric ratio of about 98:2, again corresponding to an enantiomeric excess >95% for both above mentioned HFAs.

It is worth noting that the studied *Lactobacillus* strain hydrates the three UFAs with identical stereoselectivity. The descriptor switch from (*R*) form of hydroxystearic acid to the (*S*) form of (12Z)-10-hydroxy-octadecenoic and (12Z,15Z)-10-hydroxy-octadecadienoic acid is due only to a change of substituent priority, according to the Cahn-Ingold-Prelog rules. Most likely, the oleate hydratase(s) produced by *Lactobacillus rhamnosus* accepts as substrates different unsaturated fatty acids, which must have a (*Z*) Δ^{9-10} double bond as the sole mandatory requirement. The latter catalyst(s) works with very high regio and stereoselectivity regardless of the presence of other double bonds on the fatty acid chain.

This wide substrates acceptance doesn't imply that the studied microorganism can transform the three selected fatty acid with the same kinetic ratio. Even though the substrates were hydrated with high selectivity, the microbial hydratase(s) could possess different affinity for each one of the acids that, in turn, could show different reactions kinetic.

In order to investigate this point, we set up two experiments based on the biotransformation of the mixture of the three acids. Accordingly, we added a 1:1:1 mixture of oleic/linoleic/linolenic acids to an anaerobic culture of *Lactobacillus rhamnosus*. The overall acids concentration was set at 6 g/L and the experiments were performed in triplicate using both the flask- and fermenter-based procedures. The biotransformations were stopped after 48 h and the crude products were derivatized and then analysed by GC-MS, in order to measure the HFAs relative compositions. Regardless of the transformation yields, all the experiments showed that 10-hydroxystearic acid was the most abundant HFA. Flask-based biotransformations afforded **4a**/**4b**/**4c** in a 59/17/24 ratio, whereas fermenter-based biotransformations gave **4a**/**4b**/**4c** in a 71/15/14 ratio. Overall, it seems that *Lactobacillus rhamnosus* can hydrate oleic acid faster than linoleic and linolenic acids. This could be due either to the presence of different hydratases or to the specific activity of a single hydratase towards each one of the UFAs used in this study, thus justifying the different product ratios. In spite of this fact, yields are not related to this aspect, as demonstrated by the fact that for large-scale biotransformation experiments, the hydration of linoleic acid affords hydroxy acid **4b** in yields higher than those obtained for the hydration of oleic or linolenic acid.

3. Materials and Methods

3.1. Materials and General Methods

All air- and moisture-sensitive reactions were carried out using dry solvents and under a static atmosphere of nitrogen. All solvents and reagents were of commercial quality.

Oleic acid (94%, lot. MKBZ2615V), linoleic acid (99%, lot. SLBT2627), linolenic acid (68%, lot. 310689/1), MRS broth, sodium thioglycolate, resazurin sodium salt, and glucose were purchased from Sigma-Aldrich (St. Louis, MO, USA). Linolenic acid (85% purity, lot. 81003) was purchased from Nissan—Nippon Oil and Fats Co. (Tokyo, Japan), LTD. (*S*)-*O*-acetyl mandelic acid was prepared starting from (*S*)-mandelic acid and using acetic anhydride, pyridine and cat. DMAP, as described previously [45].

A reference standard sample of 10-(*R*)-hydroxystearic acid, showing 21% ee, was prepared by baker's yeast-mediated hydration of oleic acid, according to the biotransformation procedure described in our previous work [25].

Reference standard samples of racemic (12*Z*)-10-hydroxy-octadecenoic acid and (12*Z*,15*Z*)-10-hydroxy-octadecadienoic acid were prepared starting from the corresponding (*S*) enantiomers obtained by biotransformation. The process is based on the following two steps chemical transformation.

A solution of dry DMSO (0.5 mL, 7 mmol) in CH_2Cl_2 (3 mL) was added dropwise to a stirred solution of oxalyl chloride (0.3 mL, 3.5 mmol) in CH_2Cl_2 (7 mL) at −70 °C. After ten minutes, a sample of enantio-enriched (12*Z*)-10-hydroxy-octadecenoic acid (**4b**) or (12*Z*,15*Z*)-10-hydroxy-octadecadienoic acid (**4c**) (300 mg, 1 mmol) in CH_2Cl_2 (2 mL) was added dropwise. After a further 15 min, dry Et_3N (2 mL, 14.3 mmol) was added and the resulting mixture was allowed to warm to room temperature. The reaction was then poured into ice-cooled water and was extracted twice with CH_2Cl_2 (50 mL × 2). The combined organic phases was washed with brine and concentrated under reduced pressure. The residue was dissolved in methanol (30 mL) and was treated at 0 °C with $NaBH_4$ (100 mg, 2.6 mmol) under stirring. After complete reduction of the ketone (TLC analysis), the reaction was quenched by addition of diluted HCl aq. (3% *w*/*w*, 40 mL) followed by extraction with CH_2Cl_2 (50 mL × 2). The combined organic phases was washed with brine and concentrated under reduced pressure. The residue was purified by chromatography using *n*-hexane/AcOEt (9:1–7:3) as eluent to afford the racemic hydroxy acid derivatives **4b** or **4c** (195–230 mg, 65–77% yield).

3.2. Analytical Methods and Characterization of the Products Deriving from the Biotransformation Experiments

^{1}H- and ^{13}C-NMR Spectra and DEPT experiments: $CDCl_3$ solutions at RT using a Bruker-AC-400 spectrometer (Billerica, MA, USA) at 400, 100, and 100 MHz, respectively; ^{13}C spectra are proton decoupled; chemical shifts in ppm relative to internal $SiMe_4$ (=0 ppm).

TLC: Merck silica gel *60 F_{254}* plates (Merck Millipore, Milan, Italy). Column chromatography: silica gel.

Melting points were measured on a Reichert apparatus, equipped with a Reichert microscope, and are uncorrected.

Optical rotations were measured on a Jasco-DIP-181 digital polarimeter (Tokyo, Japan).

Optical density value were measured on a Jasco V-560 UV–VIS spectrophotomer (Tokyo, Japan) at a wavelength of 600 nm.

Mass spectra were recorded on a Bruker ESQUIRE 3000 PLUS spectrometer (ESI detector) (Billerica, MA, USA) or by GC-MS analyses.

GC-MS analyses: A HP-6890 gas chromatograph equipped with a 5973 mass detector, using a *HP-5MS* column (30 m × 0.25 mm, 0.25 μm film thickness; Hewlett Packard, Palo Alto, CA, USA) was used with the following temp. program: 120° (3 min)—12°/min—195° (10 min)—12°/min—300° (10 min); carrier gas: He; constant flow 1 mL/min; split ratio: 1/30; t_R given in minutes.

The biotransformations of oleic acid, linoleic acid, and linolenic acid to give 10-hydroxystearic acid, (12Z)-10-hydroxy-octadecenoic acid, and (12Z,15Z)-10-hydroxy-octadecadienoic acid, respectively, were monitored by means of GC-MS analysis. To this end the biotransformation mixture was acidified at pH 3 and filtered on celite. The aqueous phase was then extracted three times with ethyl acetate and the combined organic layer was washed with brine and dried on Na_2SO_4. The solvent was then removed under reduced pressure and the residue was treated at 0 °C with an excess of an ethereal solution of freshly-prepared diazomethane. As soon as the evolution of nitrogen ceased, the solvent was eliminated and the residue was treated at RT with a 1:1 mixture of pyridine/acetic anhydride (4 mL for about 100 mg of residue) and DMAP (10 mg). After five hours, the excess of reagents was removed in vacuo and the residue was analysed by GC-MS in order to determine the fatty acid/hydrated fatty acid ratio.

Oleic acid methyl ester: t_R 18.95

GC-MS (EI): m/z (%) = 296 [M$^+$] (7), 264 (49), 235 (6), 222 (30), 180 (19), 166 (10), 152 (12), 137 (17), 123 (26), 110 (32), 97 (62), 83 (68), 69 (79), 55 (100).

Linoleic acid methyl ester: t_R 18.52

GC-MS (EI): m/z (%) = 294 [M$^+$] (18), 263 (15), 234 (1), 220 (4), 178 (6), 164 (10), 150 (16), 135 (15), 123 (18), 109 (36), 95 (70), 81 (93), 67 (100), 55 (56).

Linolenic acid methyl ester: t_R 18.79

GC-MS (EI): m/z (%) = 292 [M$^+$] (7), 261 (4), 249 (2), 236 (5), 191 (3), 173 (5), 149 (13), 135 (15), 121 (20), 108 (34), 95 (56), 79 (100), 67 (66), 55 (43).

Methyl 10-acetoxystearate (**8a**): t_R 24.47

GC-MS (EI): m/z (%) = 313 [M$^+$-MeCO] (6), 296 [M$^+$-AcOH] (3), 281 (17), 264 (31), 243 (11), 222 (9), 201 (100), 169 (64), 157 (16), 125 (21), 97 (18), 83 (19), 69 (21), 55 (27).

Methyl (12Z)-10-acetoxy-octadecenoate (**8b**): t_R 24.28

GC-MS (EI): m/z (%) 311 [M$^+$-MeCO] (<1), 294 [M$^+$-AcOH] (39), 279 (1), 263 (24), 220 (7), 201 (46), 169 (100), 150 (13), 136 (9), 123 (15), 109 (21), 95 (37), 81 (53), 67 (46), 55 (32).

Methyl (12Z,15Z)-10-acetoxy-octadecadienoate (**8c**): t_R 24.33

GC-MS (EI): m/z (%) 292 [M$^+$-AcOH] (76), 277 (1), 261 (20), 201 (33), 169 (100), 149 (19), 135 (28), 121 (41), 108 (42), 93 (57), 79 (87), 55 (39).

The enantiomeric composition of the isolated 10-hydroxystearic acid, (12Z)-10-hydroxy-octadecenoic acid, and (12Z,15Z)-10-hydroxy-octadecadienoic acid samples obtained from the biotransformation experiments was determined by ^{1}H-NMR analysis, according to the Rosazza procedure [43]. Hence, each one of the hydroxy acid samples (100 mg, 0.33 mmol) was treated with an excess of an ethereal solution of freshly-prepared diazomethane. As soon as the evolution of nitrogen ceased, the solvent was eliminated and the resulting methyl ester was dissolved in dry CH_2Cl_2 (5 mL) treated with (S)-O-acetylmandelic acid **9** (130 mg, 0.67 mmol), DCC (140 mg, 0.68 mmol) and DMAP (10 mg), stirring at RT for 6 h. The reaction was then quenched by the addition of water and diethyl ether (60 mL). The formed dicyclohexylurea was removed by filtration on celite and the organic phase was washed with aq. $NaHCO_3$, brine and dried on Na_2SO_4. The solvent was then removed under reduced pressure and the residue was roughly purified by chromatography, collecting every fraction containing the fatty acid mandelates.

3.3. Biotransformation Experiments

Lactobacillus rhamnosus (ATCC 53103), in lyophilized form, was purchased from Malesci Spa (Florence, Italy) (trade name Kaleidon 60). The microorganism was grown anaerobically at 37 °C, under a nitrogen atmosphere. The biotransformation experiments were performed either in flasks or in a 5 L fermenter (Biostat A BB-8822000, Sartorius-Stedim (Göttingen, Germany)) using MRS broth as the medium. Unless otherwise stated, all the biotransformation experiments were carried out in triplicate.

3.3.1. Representative Procedure for Flask-Scale Biotransformations

The anaerobic flasks were prepared loading 40 mL of MRS broth in 100 mL conical vacuum flasks followed by the addition of cysteine (20 mg), sodium thioglycolate (40 mg) and resazurine sodium salt (1 mg). The flasks were flushed with nitrogen until complete removal of the oxygen content, then were sealed with a silicone rubber septa and sterilized (121 °C, 15 min.). Each flask was inoculated via syringe with lyophilized *Lactobacillus rhamnosus* (3×10^9 CFU, suspended in 2 mL of sterilized skimmed milk) and was incubated at 37 °C and at 130 rpm.

A solution of the fatty acid (120 mg) in ethanol (0.15 mL) and 2 mL of a sterilized solution of glucose (300 g/L) in water were added to each flask after 3.5 and 6 h since the inoculum, respectively. After 48 h the reaction mixtures were acidified at pH 3 by addition of diluted HCl and then filtered on celite. The aqueous phases are then extracted three times with ethyl acetate and the combined organic layers were washed with brine, dried on Na_2SO_4 and the solvent was removed under reduced pressure. The crude biotransformation mixtures were derivatized and analysed by GC-MS as described above (Section 3.2).

3.3.2. Representative Procedure for Preparative Biotransformations

Two anaerobic flasks, containing 40 mL of MRS broth and prepared as described above, were inoculated with *Lactobacillus rhamnosus* (3×10^9 CFU for each flask) and then incubated at 37 °C and 130 rpm for 6 h. The cultures were centrifuged at $3220\times g$ for 3 min (4 °C), the supernatant removed and the cells were resuspended in 4 mL of sterilized skimmed milk. The obtained suspension was added to a fermenter vessel containing nitrogen flushed MRS broth (1 L). The temperature, the stirring speed, and the pH were set to 37 °C, 170 rpm and 6.2, respectively. The pH was controlled by dropwise addition of sterilized aqueous solutions (10% w/w in water) of either acetic acid or ammonia. About 2–4 h since the inoculum, the fermentation showed an exponential phase of growth as indicated by starting of the continuous addition of base, necessary to neutralize the lactic acid produced by the glucose bacterial catabolism. As soon as 60 mmol (6 h) and 220 mmol (10 h) of ammonia were supplemented, the solution of the suitable fatty acid (5 g) in ethanol (5 mL) and then 65 mL of a sterilized solution of glucose (300 g/L) in water were added at once. The fermentation was stopped 48 h since the inoculum. At that time the reaction mixture was acidified at pH 3 by addition of diluted HCl and then filtered on celite. The aqueous phase was then extracted three times with ethyl acetate and the combined organic layers were washed with brine, dried on Na_2SO_4, and the solvent was removed under reduced pressure. The residue was purified by chromatography using *n*-hexane/AcOEt (9:1–7:3) as the eluent to afford unreacted fatty acid (first eluted fractions) followed by hydroxy acid derivative.

The general preparative procedure was performed using oleic acid as substrate. The resulting crude biotransformation mixture showed a unreacted oleic acid/10-hydroxystearic acid ratio of 2:3 (by GC-MS analysis). The chromatographic purification allowed isolating 1.5 g of unreacted oleic acid and 2.2 g of 10-hydroxystearic acid (colorless crystal; 41% yield; 59% yield versus transformed oleic acid). A sample of the obtained 10-hydroxystearic acid was transformed in the corresponding (*S*)-*O*-acetylmandelate ester, whose NMR analysis confirmed that the hydroxy acid is the (*R*)-enantiomer possessing ee > 95%.

(R)-10-Hydroxystearic acid (**4a**): m.p.: 82–84 °C; ^{1}H-NMR (400 MHz, $CDCl_3$): δ = 3.65–3.53 (m, 1H), 2.34 (t, *J* = 7.5 Hz, 2H), 1.69–1.56 (m, 2H), 1.51–1.20 (m, 27H), 0.88 (t, *J* = 7.0 Hz, 3H). ^{13}C-NMR (100 MHz, $CDCl_3$): δ = 178.7 (C), 72.1 (CH), 37.5 (CH_2), 37.4 (CH_2), 33.8 (CH_2), 31.9 (CH_2), 29.7 (CH_2), 29.6 (CH_2), 29.6 (CH_2), 29.3 (CH_2), 29.3 (CH_2), 29.1 (CH_2), 29.0 (CH_2), 25.6 (CH_2), 25.5 (CH_2), 24.6 (CH_2), 22.6 (CH_2), 14.1 (Me). MS (ESI): 299.1 (M − 1, negative ions).

The general preparative procedure was performed using linoleic acid as substrate. The resulting crude biotransformation mixture showed a unreacted linoleic acid/(12*Z*)-10-hydroxy-octadecenoic acid ratio of 1:1 (by GC-MS analysis). The chromatographic purification allowed isolating 2 g of unreacted linoleic acid and 2.4 g of (12*Z*)-10-hydroxy-octadecenoic acid (pale yellow oil; 45% yield; 75% yield versus transformed linoleic acid). A sample of the obtained 10-hydroxy-octadecenoic acid was transformed in the corresponding (*S*)-*O*-acetylmandelate ester, whose NMR analysis indicated that the hydroxy acid possessed ee > 95%. The absolute configuration was established as (*S*) by measurement of its optical rotation value.

(S)-(12Z)-10-hydroxy-octadecenoic acid (**4b**): $[α]_D^{20}$ = −6.4 (*c* 2.6, MeOH). ^{1}H-NMR (400 MHz, $CDCl_3$) δ 5.61–5.51 (m, 1H), 5.44–5.35 (m, 1H), 3.67–3.58 (m, 1H), 2.34 (t, *J* = 7.5 Hz, 2H), 2.22 (t, *J* = 6.7 Hz, 2H), 2.09–2.00 (m, 2H), 1.68–1.58 (m, 2H), 1.52–1.22 (m, 18H), 0.89 (t, *J* = 6.9 Hz, 3H). ^{13}C-NMR (100 MHz, $CDCl_3$) δ 179.5 (C), 133.5 (CH), 125.0 (CH), 71.6 (CH), 36.7 (CH_2), 35.3 (CH_2), 34.0 (CH_2), 31.5 (CH_2), 29.5 (CH_2), 29.3 (CH_2), 29.3 (CH_2), 29.1 (CH_2), 29.0 (CH_2), 27.4 (CH_2), 25.6 (CH_2), 24.6 (CH_2), 22.5 (CH_2), 14.0 (Me). MS (ESI): 321.4 (M + Na^+); 297.2 (M − 1, negative ions).

The general preparative procedure was performed using linolenic acid as a substrate. The resulting crude biotransformation mixture showed a unreacted linolenic acid/(12*Z*,15*Z*)-10-hydroxy-octadecadienoic acid ratio of 2:1 (by GC-MS analysis). The chromatographic purification allowed isolating 2.9 g of unreacted linolenic acid and 1.8 g of (12*Z*,15*Z*)-10-hydroxy-octadecadienoic acid (pale yellow oil; 34% yield; 80% yield versus transformed linolenic acid). A sample of the obtained 10-hydroxy-octadecadienoic acid was transformed in the corresponding (*S*)-*O*-acetylmandelate ester, whose NMR analysis indicated that the hydroxy-acid possessed ee > 95%. The absolute configuration was established as (*S*) by measurement of its optical rotation value.

(S)-(12Z,15Z)-10-hydroxy-octadecadienoic acid (**4c**): $[α]_D^{20}$ = −4.7 (*c* 2.5, MeOH). ^{1}H-NMR (400 MHz, $CDCl_3$) δ 5.63–5.25 (m, 4H), 3.68–3.59 (m, 1H), 2.81 (t, *J* = 7.2 Hz, 2H), 2.34 (t, *J* = 7.5 Hz, 2H), 2.28–2.20 (m, 2H), 2.12–2.02 (m, 2H), 1.69–1.57 (m, 2H), 1.53–1.22 (m, 13H), 0.97 (t, *J* = 7.5 Hz, 3H). ^{13}C-NMR (100 MHz, $CDCl_3$) δ 179.5 (C), 132.2 (CH), 131.5 (CH), 126.8 (CH), 125.4 (CH), 71.5 (CH), 36.8 (CH_2), 35.3 (CH_2), 34.0 (CH_2), 29.5 (CH_2), 29.3 (CH_2), 29.1 (CH_2), 29.0 (CH_2), 25.7 (CH_2), 25.6 (CH_2), 24.6 (CH_2), 20.6 (CH_2), 14.2 (Me). MS (ESI): 319.4 (M + Na^+); 295.1 (M − 1, negative ions).

4. Conclusions

The probiotic bacterium *Lactobacillus rhamnosus* (ATCC 53103) can be used as a whole-cell biocatalyst for the hydration of oleic acid, linoleic acid, and linolenic acid to produce (*R*)-10-hydroxystearic acid, (*S*)-(12*Z*)-10-hydroxy-octadecenoic acid, and (*S*)-(12*Z*,15*Z*)-10-hydroxy-octadecadienoic acid, respectively. We developed a biotransformation protocol that holds preparative significance because it is scalable and allows obtaining the above-mentioned HFAs with high regio- and stereoselectivity (ee > 95%). Finally, the used bacteria strain does not involve any safety concerns and the proposed process can be employed for the industrial production of food flavour.

Acknowledgments: The authors thank Cariplo Foundation for supporting this study within the project no. 2017-1015 SOAVE (Seed and vegetable Oils Active Valorization through Enzymes).

Author Contributions: S.S. and D.D.S. equally contributed to conceive, design, and perform the experiments as well as to analyse the data. S.S. wrote the paper.

Conflicts of Interest: The authors declare no conflict of interest.

References

1. Kim, K.-R.; Oh, D.-K. Production of hydroxy fatty acids by microbial fatty acid-hydroxylation enzymes. *Biotechnol. Adv.* **2013**, *31*, 1473–1485. [CrossRef] [PubMed]
2. Cao, Y.; Zhang, X. Production of long-chain hydroxy fatty acids by microbial conversion. *Appl. Microbiol. Biotechnol.* **2013**, *97*, 3323–3331. [CrossRef] [PubMed]
3. Lu, W.; Ness, J.E.; Xie, W.; Zhang, X.; Minshull, J.; Gross, R.A. Biosynthesis of monomers for plastics from renewable oils. *J. Am. Chem. Soc.* **2010**, *132*, 15451–15455. [CrossRef] [PubMed]
4. Serra, S.; Fuganti, C.; Brenna, E. Biocatalytic preparation of natural flavours and fragrances. *Trends Biotechnol.* **2005**, *23*, 193–198. [CrossRef] [PubMed]
5. Braga, A.; Belo, I. Biotechnological production of γ-decalactone, a peach like aroma, by *Yarrowia lipolytica*. *World J. Microbiol. Biotechnol.* **2016**, *32*, 169. [CrossRef] [PubMed]
6. Waché, Y.; Aguedo, M.; Nicaud, J.-M.; Belin, J.-M. Catabolism of hydroxyacids and biotechnological production of lactones by *Yarrowia lipolytica*. *Appl. Microbiol. Biotechnol.* **2003**, *61*, 393–404. [CrossRef] [PubMed]
7. Kim, A.Y. Application of biotechnology to the production of natural flavor and fragrance chemicals. In *Natural Flavors and Fragrances*; American Chemical Society: Washington, DC, USA, 2005; Volume 908, pp. 60–75, ISBN 0-8412-3904-5.
8. Maurer, B.; Hauser, A. Identification and synthesis of new γ-lactones from tuberose absolute (*Polianthes tuberosa*). *Helv. Chim. Acta* **1982**, *65*, 462–476. [CrossRef]
9. Wallen, L.L.; Benedict, R.G.; Jackson, R.W. The microbiological production of 10-hydroxystearic acid from oleic acid. *Arch. Biochem. Biophys.* **1962**, *99*, 249–253. [CrossRef]
10. Davis, E.N.; Wallen, L.L.; Goodwin, J.C.; Rohwedder, W.K.; Rhodes, R.A. Microbial hydration of *cis*-9-alkenoic acids. *Lipids* **1969**, *4*, 356–362. [CrossRef] [PubMed]
11. Seo, C.W.; Yamada, Y.; Takada, N.; Okada, H. Hydration of squalene and oleic acid by *Corynebacterium* sp. S-401. *Agric. Biol. Chem.* **1981**, *45*, 2025–2030. [CrossRef]
12. Koritala, S.; Hou, C.T.; Hesseltine, C.W.; Bagby, M.O. Microbial conversion of oleic acid to 10-hydroxystearic acid. *Appl. Microbiol. Biotechnol.* **1989**, *32*, 299–304. [CrossRef]
13. El-Sharkawy, S.H.; Yang, W.; Dostal, L.; Rosazza, J.P. Microbial oxidation of oleic acid. *Appl. Environ. Microbiol.* **1992**, *58*, 2116–2122. [PubMed]
14. Koritala, S.; Bagby, M.O. Microbial conversion of linoleic and linolenic acids to unsaturated hydroxy fatty acids. *J. Am. Oil Chem. Soc.* **1992**, *69*, 575–578. [CrossRef]
15. Kaneshiro, T.; Huang, J.-K.; Weisleder, D.; Bagby, M.O. 10(*R*)-Hydroxystearic acid production by a novel microbe, NRRL B-14797, isolated from compost. *J. Ind. Microbiol.* **1994**, *13*, 351–355. [CrossRef]
16. Hou, C.T. Conversion of linoleic acid to 10-hydroxy-12(Z)-octadecenoic acid by *Flavobacterium* sp. (NRRL B-14859). *J. Am. Oil Chem. Soc.* **1994**, *71*, 975–978. [CrossRef]
17. Gocho, S.; Tabogami, N.; Inagaki, M.; Kawabata, C.; Komai, T. Biotransformation of oleic acid to optically active γ-dodecalactone. *Biosci. Biotechnol. Biochem.* **1995**, *59*, 1571–1572. [CrossRef]
18. Hudson, J.A.; MacKenzie, C.A.M.; Joblin, K.N. Conversion of oleic acid to 10-hydroxystearic acid by two species of ruminal bacteria. *Appl. Microbiol. Biotechnol.* **1995**, *44*, 1–6. [CrossRef] [PubMed]
19. Hou, C.T. Is strain DS5 hydratase a C-10 positional specific enzyme? Identification of bioconversion products from α- and γ-linolenic acids by *Flavobacterium* sp. DS5. *J. Ind. Microbiol.* **1995**, *14*, 31–34. [CrossRef]

20. Morvan, B.; Joblin, K.N. Hydration of oleic acid by *Enterococcus gallinarum*, *Pediococcus acidilactici* and *Lactobacillus* sp. isolated from the rumen. *Anaerobe* **1999**, *5*, 605–611. [CrossRef]

21. Kishimoto, N.; Yamamoto, I.; Toraishi, K.; Yoshioka, S.; Saito, K.; Masuda, H.; Fujita, T. Two distinct pathways for the formation of hydroxy FA from linoleic acid by lactic acid bacteria. *Lipids* **2003**, *38*, 1269–1274. [CrossRef] [PubMed]

22. Kim, M.H.; Park, M.S.; Chung, C.H.; Kim, C.T.; Kim, Y.S.; Kyung, K.H. Conversion of unsaturated food fatty acids into hydroxy fatty acids by lactic acid bacteria. *J. Microbiol. Biotechnol.* **2003**, *13*, 360–365.

23. Kim, B.-N.; Yeom, S.-J.; Oh, D.-K. Conversion of oleic acid to 10-hydroxystearic acid by whole cells of *Stenotrophomonas nitritireducens*. *Biotechnol. Lett.* **2011**, *33*, 993–997. [CrossRef] [PubMed]

24. Takeuchi, M.; Kishino, S.; Tanabe, K.; Hirata, A.; Park, S.-B.; Shimizu, S.; Ogawa, J. Hydroxy fatty acid production by *Pediococcus* sp. *Eur. J. Lipid Sci. Technol.* **2013**, *115*, 386–393. [CrossRef]

25. Serra, S.; De Simeis, D. New insights on the baker's yeast-mediated hydration of oleic acid: The bacterial contaminants of yeast are responsible for the stereoselective formation of (*R*)-10-hydroxystearic acid. *J. Appl. Microbiol.* **2018**, *124*, 719–729. [CrossRef] [PubMed]

26. Bevers, L.E.; Pinkse, M.W.H.; Verhaert, P.D.E.M.; Hagen, W.R. Oleate hydratase catalyzes the hydration of a nonactivated carbon-carbon bond. *J. Bacteriol.* **2009**, *191*, 5010–5012. [CrossRef] [PubMed]

27. Joo, Y.-C.; Jeong, K.-W.; Yeom, S.-J.; Kim, Y.-S.; Kim, Y.; Oh, D.-K. Biochemical characterization and fad-binding analysis of oleate hydratase from *Macrococcus caseolyticus*. *Biochimie* **2012**, *94*, 907–915. [CrossRef] [PubMed]

28. Yang, B.; Chen, H.; Song, Y.; Chen, Y.Q.; Zhang, H.; Chen, W. Myosin-cross-reactive antigens from four different lactic acid bacteria are fatty acid hydratases. *Biotechnol. Lett.* **2013**, *35*, 75–81. [CrossRef] [PubMed]

29. Jo, Y.-S.; An, J.-U.; Oh, D.-K. γ-dodecelactone production from safflower oil via 10-hydroxy-12(*Z*)-octadecenoic acid intermediate by whole cells of *Candida boidinii* and *Stenotrophomonas nitritireducens*. *J. Agric. Food Chem.* **2014**, *62*, 6736–6745. [CrossRef] [PubMed]

30. Oh, H.-J.; Kim, S.-U.; Song, J.-W.; Lee, J.-H.; Kang, W.-R.; Jo, Y.-S.; Kim, K.-R.; Bornscheuer, U.T.; Oh, D.-K.; Park, J.-B. Biotransformation of linoleic acid into hydroxy fatty acids and carboxylic acids using a linoleate double bond hydratase as key enzyme. *Adv. Synth. Catal.* **2015**, *357*, 408–416. [CrossRef]

31. Takeuchi, M.; Kishino, S.; Hirata, A.; Park, S.-B.; Kitamura, N.; Ogawa, J. Characterization of the linoleic acid Δ9 hydratase catalyzing the first step of polyunsaturated fatty acid saturation metabolism in *Lactobacillus plantarum* AKU 1009a. *J. Biosci. Bioeng.* **2015**, *119*, 636–641. [CrossRef] [PubMed]

32. Kim, K.-R.; Oh, H.-J.; Park, C.-S.; Hong, S.-H.; Park, J.-Y.; Oh, D.-K. Unveiling of novel regio-selective fatty acid double bond hydratases from *Lactobacillus acidophilus* involved in the selective oxyfunctionalization of mono- and di-hydroxy fatty acids. *Biotechnol. Bioeng.* **2015**, *112*, 2206–2213. [CrossRef] [PubMed]

33. Hiseni, A.; Arends, I.W.C.E.; Otten, L.G. New cofactor-independent hydration biocatalysts: Structural, biochemical, and biocatalytic characteristics of carotenoid and oleate hydratases. *ChemCatChem* **2015**, *7*, 29–37. [CrossRef]

34. Chen, Y.Y.; Liang, N.Y.; Curtis, J.M.; Gänzle, M.G. Characterization of linoleate 10-hydratase of *Lactobacillus plantarum* and novel antifungal metabolites. *Front. Microbiol.* **2016**, *7*, 1561. [CrossRef] [PubMed]

35. Kang, W.-R.; Seo, M.-J.; An, J.-U.; Shin, K.-C.; Oh, D.-K. Production of δ-decalactone from linoleic acid via 13-hydroxy-9(*Z*)-octadecenoic acid intermediate by one-pot reaction using linoleate 13-hydratase and whole *Yarrowia lipolytica* cells. *Biotechnol. Lett.* **2016**, *38*, 817–823. [CrossRef] [PubMed]

36. Demming, R.M.; Otte, K.B.; Nestl, B.M.; Hauer, B. Optimized reaction conditions enable the hydration of non-natural substrates by the oleate hydratase from *Elizabethkingia meningoseptica*. *ChemCatChem* **2017**, *9*, 758–766. [CrossRef]

37. Lorenzen, J.; Driller, R.; Waldow, A.; Qoura, F.; Loll, B.; Bruck, T. *Rhodococcus erythropolis* oleate hydratase: A new member in the oleate hydratase family tree biochemical and structural studies. *ChemCatchem* **2018**, *10*, 407–414. [CrossRef]

38. Bourdichon, F.; Casaregola, S.; Farrokh, C.; Frisvad, J.C.; Gerds, M.L.; Hammes, W.P.; Harnett, J.; Huys, G.; Laulund, S.; Ouwehand, A.; et al. Food fermentations: Microorganisms with technological beneficial use. *Int. J. Food Microbiol.* **2012**, *154*, 87–97. [CrossRef] [PubMed]

39. Gorbach, S.H.; Goldin, B.R. *Lactobacillus* Strains and Methods of Selection. U.S. Patent 4839281, 13 June 1989.

40. Ku, S.; You, H.J.; Park, M.S.; Ji, G.E. Whole-cell biocatalysis for producing ginsenoside Rd from Rb1 using *Lactobacillus rhamnosus* GG. *J. Microbiol. Biotechnol.* **2016**, *26*, 1206–1215. [CrossRef] [PubMed]

41. Sturme, M.H.J.; Francke, C.; Siezen, R.J.; de Vos, W.M.; Kleerebezem, M. Making sense of quorum sensing in lactobacilli: A special focus on *Lactobacillus plantarum* WCSF1. *Microbiology* **2007**, *153*, 3939–3947. [CrossRef] [PubMed]

42. Murata, A.; Kai, K.; Tsutsui, K.; Takeuchi, J.; Todoroki, Y.; Furihata, K.; Yokoyama, M.; Baldermann, S.; Watanabe, N. Enantio-selective reduction of the flowering related compound KODA and its analogues in *Pharbitis nil* cv. Violet. *Tetrahedron* **2012**, *68*, 5583–5589. [CrossRef]

43. Yang, W.; Dostal, L.; Rosazza, J.P.N. Stereospecificity of microbial hydrations of oleic acid to 10-hydroxystearic acid. *Appl. Environ. Microbiol.* **1993**, *59*, 281–284. [PubMed]

44. Mancuso, A.J.; Huang, S.-L.; Swern, D. Oxidation of long-chain and related alcohols to carbonyls by dimethyl sulfoxide "activated" by oxalyl chloride. *J. Org. Chem.* **1978**, *43*, 2480–2482. [CrossRef]

45. Ebbers, E.J.; Ariaans, G.J.A.; Bruggink, A.; Zwanenburg, B. Controlled racemization and asymmetric transformation of α-substituted carboxylic acids in the melt. *Tetrahedron Asymmetry* **1999**, *10*, 3701–3718. [CrossRef]

MDPI

Article

Microbial Kinetic Resolution of Aroma Compounds Using Solid-State Fermentation

Filip Boratyński * [ID], Ewa Szczepańska, Aleksandra Grudniewska and Teresa Olejniczak

Department of Chemistry, Wroclaw University of Environmental and Life Sciences, Norwida 25, 50-375 Wrocław, Poland; ewa.szczepanska@upwr.edu.pl (E.S.); aleksandra.grudniewska@upwr.edu.pl (A.G.); teresa.olejniczak@upwr.edu.pl (T.O.)
* Correspondence: filip.boratynski@upwr.edu.pl; Tel.: +48-71-320-5257

Received: 14 December 2017; Accepted: 13 January 2018; Published: 16 January 2018

Abstract: A novel microbial approach to the production of enantiomerically enriched and pure aroma compounds based on kinetic resolution via solid-state fermentation is proposed. Twenty-five filamentous fungi were screened for lipase activity and enantioselective hydrolysis of a volatile racemic ester (1-phenylethyl acetate (**1**)) and several racemic lactones (*trans* and *cis* whisky lactones (**4, 5**), γ-decalactone (**7**), δ-decalactone (**8**), (*cis*-3a,4,7,7a-tetrahydro-1(3*H*)-isobenzofuranone (**9**)). Solid-state fermentation was conducted with linseed and rapeseed cakes. Kinetic resolution afforded enantiomerically enriched products with high enantiomeric excesses (ee = 82–99%). The results highlight the potential economic value of solid-state fermentation using agroindustrial side-stream feedstocks as an alternative to more expensive processes conducted in submerged fermentation.

Keywords: aroma compounds; kinetic resolution; solid-state fermentation; agro-industrial side stream; rapeseed cake; linseed cake; lactones; esters

1. Introduction

The food industry generates large quantities of wastes and by-products, and research interest in efficient use of agroindustrial residues has been increasing [1]. Several bioprocesses that use these residues as substrates have been developed, including production of enzymes, single cell proteins, ethanol, organic acids, biopolymers, and secondary metabolites [2]. Solid-state fermentation (SSF) constitutes a microbial culture method alternative to submerged fermentation (SmF). SSF lowers the capital investment that is required for particular bioprocesses by approximately 78% in comparison to SmF [3]. Because growth media account for approximately 40% of the total cost of bioprocessing, it is reasonable to use inexpensive raw materials such, as agricultural by-products [4].

Oilseed cakes are solid residues that are obtained after pressing of oil seeds. They can constitute up to 75% of total seed weight. Oilseed cakes are rich in carbohydrates, proteins, fat, and cellulose, and therefore provide excellent media for growth of microorganisms [5]. The world market for oilseed cakes is dominated by soybean, rapeseed, cottonseed, groundnut, sunflower, and linseed cakes.

Microbial SSF on renewable agroindustrial side-stream products is ideal for efficient production of industrially important biocatalysts, such as lipases, proteases, cellulases, and amylases [6]. Application of microbial enzymes or whole cells permits transformation with high chemo-, regio-, and enantioselectivity [7]. Notably, biotransformation is an environmentally friendly process because it can be conducted under mild conditions, requires few chemicals, and produces little toxic chemical waste. In the pharmaceutical, agricultural, and fine chemical industries, there is a strong demand for the production of the enantiopure forms of chiral compounds, and biocatalysis is therefore being used to manufacture a wide range of products [8].

Increasing attention is being paid to the origins of food additives, and those with natural origins are preferred. Compounds obtained by biotransformation, according to United States and European Union regulations, are regarded as natural [9]. Interest in biotechnological production of natural and natural-identical flavor compounds has recently increased [10]. One group of additives with well-characterized flavor properties are compounds that contain ester bonds, including lactones, which are characterized by an intense, specific aroma, and which are used in the food, cosmetic, and pharmaceutical industries. Their fragrance depends on the size of the ring, the type of substituents, the presence of unsaturated bonds, and the configuration of the chiral centers [11].

One common aroma lactone is whisky lactone. It is essential for the flavoring of aged alcoholic beverages, such as whisky, cognac, brandy, and wine. It is also used as an aroma ingredient of flavored sweets and beverages, as well as a variety of baked foods and tobacco. Four stereoisomers of whisky lactone are known, and their olfactory properties are determined by their spatial structure [12]. γ-Decalactone was originally isolated from fruits, meat, and dairy products. It enriches food products with an intense scent of peach or coconut. The *S* enantiomer of γ-decalactone occurs naturally in mango, while the *R* enantiomer is found in most fruits, especially in peaches [13]. A reliable process based on the microbial transformation of castor oil secures the production of natural (+)-(*R*)-γ-decalactone, whilst the (*S*)-enantiomer is not easily available yet [14]. δ-Decalactone, with its creamy, sweet, milky, coconut-peach flavor, is of great interest to the food industry. It is a well-known constituent of the aroma of dairy products and some fruits [15]. *cis*-3a,4,7,7a-Tetrahydro-1(3*H*)-isobenzofuranone is the precursor of the phthalide-derived lactones that are abundant in fruits in the family *Apiaceae*, which are characterized by a celery-like aroma [16]. 1-Phenylethyl acetate is a well-known flavoring used in many countries as a food additive. Its aroma has been described as sweet and fruity, woody, and tropical with floral nuances; it is found in a wide range of fruits and vegetables, such as strawberry, melon, avocado, pineapple, and banana. It is a highly valued natural volatile ester that is widely used as an additive in cosmetics, foods, and pharmaceuticals [17].

The aim of this study was microbial kinetic resolution of aroma compounds. Whole cells of filamentous fungi growing on rapeseed (RC) and linseed cakes (LC) were screened for their ability to produce enantiomerically pure lactones and esters. To the best of our knowledge, only a few reports have been published on biotransformation via SSF. Although numerous hydrolases are produced by SSF, only a few reports have discussed their application in biotransformation. Macedo et al. [18] described the production of lipases by SSF and preparation of lyophilized powder of extracted enzymes used for synthesis of short chain citronellyl esters. Only one study, conducted by Nagy et al. [19], has examined lipases produced by SSF as catalysts for kinetic resolution of racemic secondary alcohols. When considering the prevalence of application of these aroma compounds in the food industry, and the economic benefits of sustainable management of agricultural side streams, this approach is undeniably attractive.

2. Results and Discussion

The stereoselectivity of hydrolysis catalyzed by enzymes produced by filamentous fungi in solid-state fermentation (SSF) was tested with various aroma compounds: 1-phenylethyl acetate (**1**), a mixture of *trans* and *cis* whisky lactones (**4**, **5**), γ-decalactone (**7**), δ-decalactone (**8**), and *cis*-3a,4,7,7a-tetrahydro-1(3*H*)-isobenzofuranone (**9**) (Scheme 1). SSF was conducted on oilseed cakes from linseeds and rapeseeds, which are by-products of the oleoindustry that contain all essential ingredients for fungal growth, especially the fatty acids that are required for lipase biosynthesis.

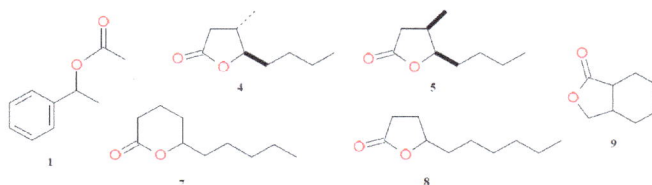

Scheme 1. Aroma compounds applied for kinetic resolution by solid-state fermentation (SSF).

2.1. Kinetic Resolution of 1-Phenylethyl Acetate (**1**)

Fungal strains exhibiting lipolytic activity were tested on racemic 1-phenylethyl acetate (**1**) as a reference substrate to investigate their capacity for dynamic kinetic resolution (Scheme 2). This substrate was hydrolyzed to phenylethanol (**2**), which is, especially in its enantiomerically pure forms, in great demand in the agrochemical, flavor, and pharmaceutical industries [20]. Additionally, as a result of alcohol **2** oxidation, acetophenone (**3**) was formed. In the biotransformation of racemic 1-phenylethyl acetate (**1**), both hydrolases and oxidoreductases play a crucial role. On the basis of gas chromatography (GC) analysis, arising acetophenone (**3**) during the biotransformation suggests that (*S*)-1-phenylethanol (**2**) is oxidized to acetophenone (**3**), which is immediately selectively reduced to (*R*)-1-phenylethanol (**2**).

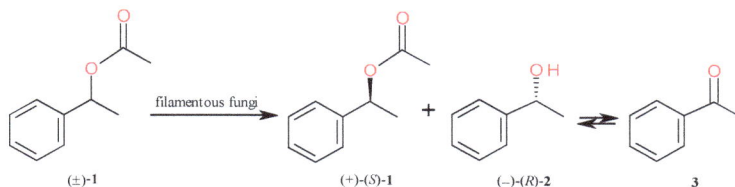

Scheme 2. Kinetic resolution of 1-phenylethyl acetate (**1**).

The majority of the biocatalysts preferentially transformed the (*R*)-**1** ester to the corresponding (*R*)-**2** alcohol (Table 1). Kinetic resolution of ester **1** catalyzed by all nine strains (*Aspergillus candidus* AM386, *A. nidulans* AM243, *Botrytis cirenea* AM235, *Fusarium oxysporum* AM21, *F. semitectum* AM20, *Mucor spinosus* AM398, *Papularia rosea* AM17, *Penicillum camemberti* AM83, *Poria placenta* AM38) of filamentous fungi tested produced (*S*)-**1** and (*R*)-**2**, with a high conversion rate and enantiomeric excess. The greatest quantity of enantiomerically pure (*S*)-**1** (ee > 99%) was achieved after three days of hydrolysis catalyzed by *F. oxysporum* AM21. Three strains (*A. nidulans* AM243, *P. camemberti* AM83, and *F. avenaceum* AM11) produced the (*R*)-**2** enantiomer (ee > 99%). Hydrolysis on RC proceeded relatively fast, and the highest enantioselectivity was achieved after three days of biotransformation. Subsequently, enantiomeric excess of product (*R*)-**2** decreased. With application of *B. cirenea* AM235 and *P. rosea* AM17, high enantiomeric excess was observed, even after 10 days; however, in *P. rosea* AM17 culture the amount of acetophenone (**3**) increased steadily with time (data not shown).

LC was also a valuable biotransformation medium; however, hydrolysis proceeded via different pathways (Table 2). Hydrolysis of **1** with *B. cirenea* AM235 on LC provided the (*S*)-enantiomer of substrate **1**, whereas on RC, the (*R*)-enantiomer of product **2** was produced (Table 1). Culture of *A. ochraceus* AM370 and *Penicillium thomi* AM91 produced only ketone **3**, which suggests high oxidoreductase activity of these strains. Biotransformation catalyzed by *Fusarium avenaceum* AM11 also produced different results, with only ketone **3** or alcohol **2** observed on RC and LC, respectively. It is worthy of mention that the type of oilseed cake influenced enzyme activity and specificity. However, the use of another medium had no or only a slight influence on the kinetic resolution process catalyzed by *A. nidulans* AM243, *A. ochraceus* AM370, *M. spinosus* AM398, *P. rosea* AM17 in comparison to RC (Table 1).

Table 1. Kinetic resolution of racemic 1-phenylethyl acetate (**1**) by fungi in rapeseed cake (in % according to GC).

Strain	Time (Days)	Lipase Activity (U/g)	Conversion (%)	(S)-1		(R)-2		3
				(%)	ee (%)	(%)	ee (%)	(%)
Aspergillus candidus AM386	3	40.4	100	0	-	86	>99	14
Aspergillus nidulans AM243	3	106.6	100	0	-	100	>99	0
Aspergillus ochraceus AM370	3	21	100	0	-	0	-	100
Aspergillus ochraceus AM456	3	90.3	88	12	74	88	22	0
Aspergillus wenthi AM413	3	96.8	30	70	30	26	64	4
Botrytis cirenea AM235	6	144.7	100	0	-	96	98	4
Fusarium avenaceum AM11	3	46	100	0	-	0	-	100
Fusarium oxysporum AM21	3	146.5	18	82	>99	15	44	3
Fusarium semitectum AM20	3	105.5	96	4	>99 [a]	56	>99	40
Fusarium tricinctum AM16	3	67	25	75	20	25	66	0
Mucor spinosus AM398	3	54.1	94	6	66 [a]	78	>99	16
Papularia rosea AM17	6	160.7	97	3	0	88	96	9
Penicillum camemberti AM83	3	144.3	100	0	-	100	>99	0
Penicillium chrysogenum AM112	6	105.1	100	0	-	100	62	0
Penicillium thomi AM91	6	155.5	98	2	>99 [a]	60	54	38
Poria placenta AM38	10	22.1	98	2	>99 [a]	84	92	14
Spicoria divaricata AM423	3	89.9	93	7	>99	59	0	34

[a] The reaction proceeded with opposite enantiomer selectivity ((*R*)-**1**).

Table 2. Kinetic resolution of racemic 1-phenylethyl acetate (**1**) by fungi in linseed cake (in % according to GC).

Strain	Time (Days)	Lipase Activity (U/g)	Conversion (%)	(S)-1		(R)-2		3
				(%)	ee (%)	(%)	ee (%)	(%)
Aspergillus nidulans AM243	3	106.6	100	0	-	100	>99	0
Aspergillus ochraceus AM370	3	21	100	0	-	0	-	100
Aspergillus ochraceus AM456	3	90.3	83	17	40	83	30	0
Aspergillus wenthi AM413	10	237.5	90	10	24	78	0	12
Botrytis cirenea AM235	3	14	10	90	80	0	-	10
Fusarium avenaceum AM11	3	46	100	0	-	100	>99	0
Fusarium oxysporum AM21	10	26.6	75	25	62	59	0	16
Fusarium semitectum AM20	3	105.5	70	30	72	34	38	36
Fusarium tricinctum AM16	3	67	44	56	26	22	86	22
Mucor spinosus AM398	3	54.1	100	0	0	95	90	5
Papularia rosea AM17	6	160.7	100	0	0	46	86	54
Penicillum camemberti AM83	3	144.3	30	70	30	30	58	0
Penicillium chrysogenum AM112	6	105.1	92	8	>99	67	50	25
Penicillium notatum AM904	6	36.5	100	0	0	92	0	8
Penicillium thomi AM91	3	39.4	100	0	-	0	-	100
Sclerophoma pythiophila AR55	3	5.6	48	52	52	35	20	13
Spicoria divaricata AM423	3	20.1	100	0	-	85	26	15

Both media, RC and LC, were effective for lipase production. In general, higher lipase activity was observed in most cultures grown on RC, however three strains *A. ochraceus* AM370, *A. wenthi* AM413, and *F. avenaceum* AM11 exhibited higher lipase activity on LC. Previous reports describe a few examples of use of these oilseed cakes as a medium for hydrolase production [6,21–25]. When considering the differences in the kinetic resolution results, it appears that the chemical composition of the media induces the production of enzymes with different enantioselectivity. Fermentation on LC produced a greater content of carbohydrates and proteins as compared to RC, although the quantity of residual oil was comparable (12–13%) [26–28]. The main differences in LC and RC are in fatty acid composition, which can strongly affect lipase biosynthesis. Both LC and RC contain a significant majority of unsaturated fatty acids (90–94%). However, in LC, α-linolenic acid predominates (50–55%), whereas RC primarily contains oleic acid (~60%), and only 1% α-linolenic acid. Oilseed cake from flax contains a similar linoleic acid content (~20%) to RC [29,30]. Moreover, the physical properties of LC, which shows significantly stronger adsorption of water (used to add moisture in SSF) than RC, might explain the differences in the efficiency of the fungal kinetic resolution process [31]. It is worth mentioning that during the SSF processes that were conducted in this experiment, not only hydrolases were produced. Acetophenone (**3**) was synthesized by oxidoreductases, which have not been assessed in SSF to date.

A few examples of fungal kinetic resolution, although in submerged fermentation, of ester **1** can be found in recent literature [32,33]. *Fusarium proliferatum* NCIM1105 used for hydrolytic kinetic resolution of racemic **1** afforded 100% enantiomerically pure (*R*)-**2** within 36 h [33]. In the present study, comparable results were obtained with *A. nidulans* AM243, *P. camemberti* AM83, and *F. avenaceum* AM11 (Tables 1 and 2). Filamentous fungi *Aspergillus flavus* CECT20475 was applied in kinetic resolution of **2**. Within 24 h at 40 °C, (*R*)-**2** was esterified into (*R*)-**1** with ee = 94.6%, and ee = 99% of (*S*)-**2** was received [34]. Nagy et al. screened filamentous fungi under SSF for lipase activity and enantioselectivity relative to **1**, using wheat bran as a medium [19]. Of the 26 fungal strains that were tested, 18 were able to provide (*R*)-**2** with high enantiomeric excess (ee > 88%). Six strains hydrolyzed **1** to produce (*S*)-**1** with ee > 83%. Enantiomerically pure (*R*)-**2** was obtained with *Chaetomium elatum* UAMH2672 and *Scopulariopsis brevicaulis* WFPL248A as a biocatalysts within 24 h, and (*S*)-**1** was received within 120 h with ee > 99% by *Chaetomium globosum* OKI270 and *Gliocladium vermoesenii* NRRL1752. Resolution of racemic **1** by *Candida antarctica* lipase B was examined by Fan et al. [35]. As a result, racemic **1** was hydrolyzed with the conversion rate 41.2% and enantiomerically pure (*S*)-**2** was obtained. Liang et al. [36] applied esterase from *Bacillus* sp. SCSIO 15121 to obtain (*R*)-**2** with the conversion rate 49%. Application of the enzymatic system of vegetables to the kinetic resolution of **1** was reported by Vanderberghe et al. [37]. The highest ee of hydrolysis products was obtained by using beetroot as a biocatalyst (ee of (*R*)-**2** 66%, (*S*)-**1** >99%). Purified microbial GDSL lipase MT6 conduct hydrolysis of **1** within 12 h and generated (*S*)-**2** with ee = 97%, however the conversion rate was 28% [38].

2.2. Kinetic Resolution of trans and cis Whisky Lactones (**4**, **5**)

SSF has not previously been applied to obtain enantiomerically pure whisky lactone, one of the most commonly used flavors in the food industry. Each enantiomer displays different biological activity, due to its structural characteristics. Therefore, it is important to evaluate the natural and economical methods of production of enantiomerically pure forms of whisky lactone. Currently, a few biotechnological methods that produce enantiomerically pure *cis* and *trans* isomers of whisky lactone are known. *Trans*-(−)-(4*R*,5*S*)- and *cis*-(+)-(4*R*,5*R*)-whisky lactones can be produced by enantioselective oxidation of diols using alcohol dehydrogenases, mainly horse liver alcohol dehydrogense (HLADH) as biocatalysts. Thereafter, (−)-(4*R*,5*S*)-isomer with yield = 51% and ee = 34%, as well as (+)-(4*R*,5*R*)-isomer with yield = 48% and ee = 64% were formed after 24 h. In the same study, filamentous fungi were used to catalyze lactonization of γ-oxoacids. Transformation catalyzed by *Beauveria bassiana* AM278 provide after 48 h *trans*-(+)-(4*S*,5*R*)-whisky lactone (yield = 55% and ee > 99%) and *cis*-(−)-(4*S*,5*S*) (yield = 45% and ee = 77%) [39].

Kinetic resolution of a diastereoisomeric mixture of whisky lactones was conducted on RC and LC (Scheme 3), similar to the previous substrate **1**. Filamentous fungi mainly catalyzed hydrolysis of (−)-(4*R*,5*S*)-**4** and (−)-(4*S*,5*S*)-**5** to the corresponding hydroxyacid **6**, leaving (+)-(4*S*,5*R*)-**4** and (+)-(4*R*,5*R*)-**5** predominantly.

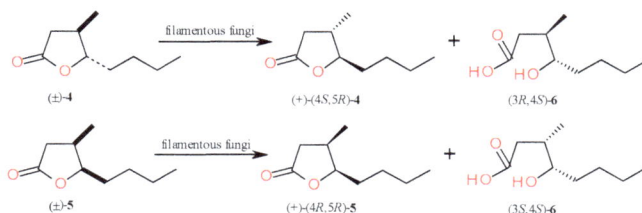

Scheme 3. Kinetic resolution of diastereoisomeric mixture of whisky lactones (**4**, **5**).

Tables 3 and 4 list the fungi that most effectively hydrolyzed **4** and **5** on LC and RC. The selected filamentous fungi exhibited biocatalytic ability to hydrolyze the internal ester bond in *trans* and *cis* whisky lactones.

Table 3. Kinetic resolution of mixture of racemic *trans* and *cis* whisky lactones (**4, 5**) by filamentus fungi in rapeseed cake (in % according to GC).

Strain	Time (Days)	Lipase Activity (U/g)	*trans/cis* Ratio (%)	*trans*-(+)-(4S,5R)-4 ee (%)	*cis*-(+)-(4R,5R)-5 ee (%)
Aspergillus sp. AM31	6	18.1	59/41	38	28
Fusarium culmorum AM9	3	3	48/52	12	14
Fusarium equiseti AM15	3	5.8	50/50	24 [a]	18
Fusarium oxysporum AM13	6	25	56/44	56	60
Papularia rosea AM17	6	160.7	33/67	70	42
Penicillum camemberti AM83	3	144.3	77/23	32	0
Penicillium chrysogenum AM112	10	49.2	77/23	42	14
Penicillium notatum AM904	6	36.5	58/42	52	12
Pycnidiella resinae AR50	6	19.8	27/73	40	0

[a] The reaction proceeded with opposite enantiomer selectivity ((−)-(4R,5S)-4).

Table 4. Kinetic resolution of mixture of racemic *trans* and *cis* whisky lactones (**4, 5**) by fungi in linseed cake (in % according to GC).

Strain	Time (Days)	Lipase Activity (U/g)	*trans/cis* Ratio (%)	*trans*-(+)-(4S,5R)-4 ee (%)	*cis*-(+)-(4R,5R)-5 ee (%)
Aspergillus nidulans AM243	6	47.2	46/54	90	0
Aspergillus ochraceus AM456	6	15.4	65/35	0	14
Fusarium avenaceum AM11	6	61.8	52/48	90	14
Fusarium semitectum AM20	6	87	40/60	26	38
Fusarium solani AM203	6	87.4	54/46	90	52
Penicillum camembertii AM83	6	118	35/65	84	0
Penicillium chrysogenum AM112	6	105.1	54/46	44	0
Penicillium notatum AM904	6	36.5	65/35	50	0
Penicillium vermiculatum AM30	10	3.2	55/45	40	10
Sclerophoma pythiophila AM55	6	9.9	56/44	28	12

Following fungal kinetic resolution of diastereoisomeric mixtures of whisky lactones (**4** and **5**), enantiomerically enriched isomers (+)-(4S,5R)-**4**, and (+)-(4R,5R)-**5** were obtained. Most of the filamentous fungi exhibited a strong tendency to hydrolyze both diastereoisomers of whisky lactone. However, a greater enantiomeric excess of (+)-(4S,5R)-**4** was observed on LC. After six days of SSF, *A. nidulans* AM243, *F. avenaceum* AM11, and *F. solani* AM203 afforded **4** with ee = 90% (Table 4). Notably, SSF with *F. solani* AM203 afforded enantiomerically enriched both diastereoisomers (ee = 90% of **4** and ee = 52% of **5**). This strain showed the highest enantioselectivity for both isomers among all of the screened strains. Further studies of medium optimization for *F. solani* AM203 will be undertaken in the near future. Biotransformation on RC was characterized by lower enantioselectivity in comparison to LC. As biotransformation progressed, kinetic resolution of **4** and **5** did not improve. The best results were achieved by *F. oxysporum* AM13 and *P. rosea* AM17, which hydrolyzed **4** with ee = 56% and 70% and **5** with ee = 60% and 42%, respectively.

2.3. Kinetic Resolution of γ-Decalactone (7) and δ-Decalactone (8)

During SSF of γ-decalactone (**7**) and δ-decalactone (**8**), both of the substrates were metabolized and probably assimilated by the microorganisms as a source of energy. Therefore, application of oilseed cake as a medium for enantioselective hydrolysis of these lactones is not reasonable. A method for obtaining enantiomerically enriched γ- and δ-decalactones by applying alcohol dehydrogenases to enzymatic oxidation of diols was presented by us previously [40]. Enzyme HLADH after two days catalyzed the oxidation of diol to the (–)-(S)-isomer of γ-decalactone with yield = 79% and ee = 20%, while PADH III after five days mediated oxidation gave (+)-(*R*)-isomer of γ-decalactone with significantly higher enantiomeric excess, but lower yield (yield = 16% and ee = 80%).

2.4. Kinetic Resolution of cis-3a,4,7,7a-Tetrahydro-1(3H)-isobenzofuranone (**9**)

Fungal kinetic resolution of **9** in SSF on RC and LC was studied. All of the strains hydrolyzed only (+)-(3a*S*,7a*R*)-**9**, whereby only (−)-(3a*R*,7a*S*)-**9** was obtained in high enantiomeric excess (Scheme 4). Table 5 shows the results of hydrolysis of **9**.

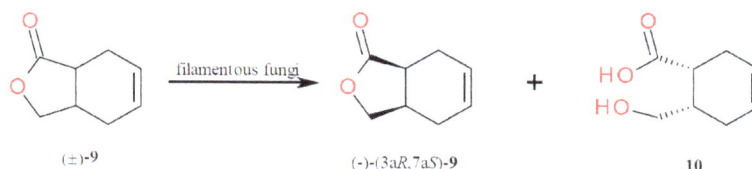

Scheme 4. Enantioselective hydrolysis of lactone **9**.

Table 5. Kinetic resolution of racemic lactone **9** by fungi in SSF (in % according to GC).

Strain	Time (Days)	Lipase Activity (U/g)	(−)-(3a*R*,7a*S*)-9 ee (%)	
			RC	LC
Aspergillus nidulans AM243	3	106.6	34	26
Aspergillus wenthi AM413	3	96.8	50	0
Botrytis cirenea AM235	6	144.7	0	80
Fusarium avenaceum AM11	3	46	10	20
Fusarium oxysporum AM21	3	146.5	74	16
Fusarium semitectum AM20	3	105.5	66	44
Fusarium tricinctum AM16	3	67	12	68
Mucor spinosus AM398	3	54.1	38	40
Papularia rosea AM17	6	160.7	26	10
Penicillum camembertii AM83	3	144.3	34	34
Penicillium chrysogenum AM112	6	105.1	36	50
Penicillium notatum AM904	6	36.5	40	82
Sclerophoma pythiophila AM55	3	5.6	20	24
Spicoria divaricata AM423	3	20.1	26	0

The highest enantioselectivity of hydrolysis of **9** was observed using *P. notatum* AM904 and *B. cirenea* AM235 as biocatalysts (ee = 82% and 80%, respectively) on LC. However, a modest enantiomeric excess of the (−)-(3a*R*,7a*S*)-9 isomer was also obtained by *F. oxysporum* AM21 and *F. semitectum* AM20 on RC (ee = 74% and 66%, respectively). Using the previous substrates (1, 4), the application of LC or RC resulted also in significant differences in hydrolysis of **9**. *B. cinerea* AM235 stereoselectively catalyzed hydrolysis of **9** only on LC, producing the (−) isomer in high enantiomeric excess (ee = 80%). *F. oxysporum* AM21 afforded the (−) isomer of **9** with ee = 74% on RC, in comparison to LC, where the enantiomeric excess was only 16%. In an alternative method applied by our group, based on bacterial oxidation of the corresponding diol, enantiomerically enriched (−)-(3a*R*,7a*S*)-9 isomer was produced by *Micrococcus* sp. DSM 30771 after seven days with the yield = 28% in comparable enantiomeric excess (ee = 88%) [41].

3. Materials and Methods

3.1. Materials

Rapeseed and linseed cakes were purchased from Oleofarm, Wroclaw, Poland. 1-Phenylethyl acetate (**1**), *trans* and *cis* whisky lactones (**4**, **5**), γ-decalactone (**7**), δ-decalactone (**8**), and *p*-nitrophenyl palmitate (*p*-NPP) were purchased from Sigma-Aldrich Chemical Co. (St. Louis, MO, USA). Racemic lactone *cis*-3a,4,7,7a-tetrahydro-1(3*H*)-isobenzofuranone (**9**) was synthesized according to described procedure [42].

3.2. Microorganisms

The following filamentous fungi strains were used for screening: *Aspergillus* sp. AM31, *Aspergillus candidus* AM386, *Aspergillus nidulans* AM243, *Aspergillus ochraceus* AM370, *Aspergillus ochraceus* AM456, *Aspergillus wenthi* AM413, *Botrytis cirenea* AM235, *Fusarium avenaceum* AM11, *Fusarium culmorum* AM9, *Fusarium equiseti* AM15, *Fusarium oxysporum* AM13, *Fusarium oxysporum* AM21, *Fusarium semitectum* AM20, *Fusarium tricinctum* AM16, *Fusarium solani* AM203, *Mucor spinosus* AM398, *Papularia rosea* AM17, *Penicillum camembertii* AM83, *Penicillium chrysogenum* AM112, *Penicillium notatum* AM904, *Penicillium thomi* AM91, *Penicillium vermiculatum* AM30, *Poria placenta* AM38, *Pycnidiella resinae* AR50, *Sclerophoma pythiophila* AR55, *Spicoria divaricata* AM423. The microorganisms were purchased from Department of Chemistry, Wroclaw University of Environmental and Life Sciences (Wroclaw, Poland). They were stored at 4 °C on Sabouraud agar slants containing peptone (10 g), glucose (30 g) and agar (15 g) dissolved in water (1 L) at pH 5.5.

3.3. Solid-State Fermentation

Oilseed cakes were placed (5 g each) in Erlenmayer flasks and autoclaved for 15 min at 121 °C. Then, hydrated to 60% moisture, inoculated with 0.5 mL of a dense spore suspension 2.3×10^7 spores/mL prepared in sterile water from agar slant cultures, and thoroughly mixed. Flasks were then incubated in thermostatic cabinet at 30 °C with defined humidity and without shaking.

3.4. Enzyme Extraction and Activity Assay

Samples (3 g) of solid-state media were taken at specified interval of cultivation time (3, 6, and 10 days), then vortexed for 5 min at 3500 rpm in phosphate buffer pH 7.2, centrifuged at 10,000 rpm for 10 min at room temperature, and supernatants were assayed for lipase activity. Lipase activity was determined in a spectrophotometric assay with *p*-NPP as a substrate. The enzyme reaction mixture contained 75 μL of substrate (1 mM) dissolved in isopropanol, 50 μL of crude enzyme filled to 3 mL by 50 mM Tris-HCl buffer (pH 8) and incubated at 37 °C for 10 min. The reaction was interrupted by addition of 1 mL cooled ethanol. The activity was measured at 410 nm. One enzyme unit (U) was defined as an amount of enzyme that released 1 μM *p*-nitrophenol per minute. Lipase activity was calculated using *p*-nitrophenol standard curve and was expressed in units/gram of oilseed cake.

3.5. Biotransformation Process

After three days of cultivation, grown cultures were sprayed by a 0.2 mL 5 mM solution of substrates in acetone and water (1:1 *v*/*v*). For each biotransformation three individual flasks were set up to estimate the progress of reaction after 3, 6, and 10 days. To the samples distilled water (15 mL) and ethyl acetate (5 mL) were added. Media were vortexed for 5 min at 3500 rpm and centrifuged at 5000 rpm for 15 min at room temperature. Finally, the organic phase was dehydrated by anhydrous MgSO$_4$ and transferred to a vial then analyzed on a gas GC instrument equipped with an autosampler (Figure 1). In control experiments, the substrates were incubated in sterile oilseed cakes without microorganism to check substrate stability. Additionally, to estimate the fungal metabolites, a control culture was performed without substrates.

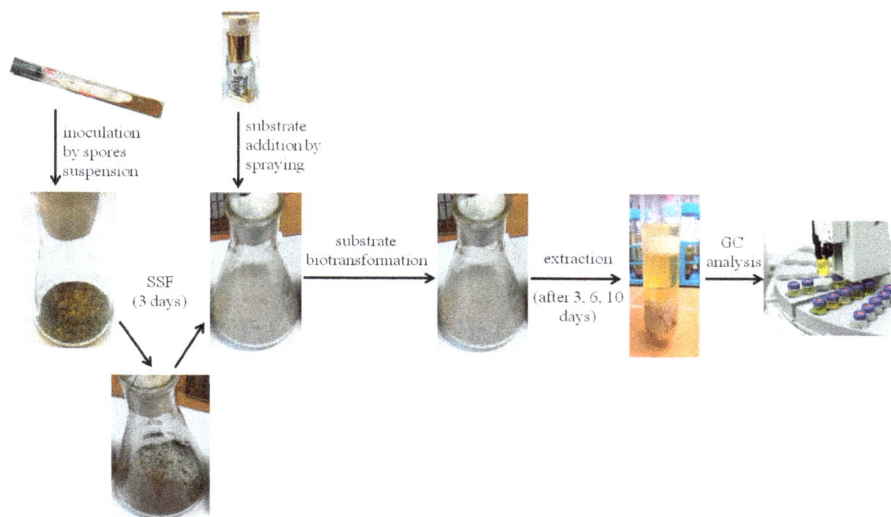

Figure 1. The process of aroma compounds kinetic resolution by using solid-state fermentation.

3.6. Analysis

The progress of reaction and enantiomeric excesses of the hydrolysis products were determined by gas chromatography. Determination of the individual isomers was based on previously obtained standards of chiral lactones [39–41]. Quantification was made when comparing with standard graph drawn for individual compounds. Gas chromatography analysis (FID, carrier gas H_2) was carried out on Agilent Technologies 7890N (GC System, Agilent, Santa Clara, CA, USA). Enantiomeric excesses of the products **1–7** and **9** were determined on chiral column Cyclosil-B (30 m × 0.25 mm × 0.25 μm), according the next temperature programs: (**1–3**) 80 °C, 110 °C (1 °C/min), 200 °C (20 °C/min) (5 min). The total run time was 39.5 min. Retention times were established, as follow: (*S*)-**1** = 28.1 min, (*R*)-**1** = 29.5 min, (*R*)-**2** = 29.9 min, (*S*)-**2** = 30.8 min, **3** = 27.5 min; (**4**, **5**) 80 °C, 160 °C (3 °C/min), 250 °C (20 °C/min) (3 min). The total run time was 34.0 min. Retention times were established as follow: (+)-(4*S*,5*R*)-**4** = 21.70 min., (−)-(4*R*,5*S*)-**4** = 22.01 min, (−)-(4*S*,5*S*)-**5** = 23.45 min, (+)-(4*R*,5*R*)-**5** = 23.60 min; (**7**) 80 °C, 210 °C (8 °C/min), 250 °C (20 °C/min) (3 min). The total run time was 21.0 min. Retention times were established as follow: (*S*)-**7** = 14.53 min, (*R*)-**7** = 14.58 min; (**9**) 80 °C, 160 °C (2 °C/min), 200 °C (20 °C/min) (6 min). The total runtime was 48.0 min. Retention times were established as follow: (+)-(3a*S*,7a*R*)-**9** = 36.3 min, (−)-(3a*R*,7a*S*)-**9** = 36.6 min. Column CP-Chirasil DEX 7502 (30 m × 0.25 mm × 0.25 μm) was used to determine **8**, according to the next temperature program: 80 °C, 120 °C (0.5 °C/min), 200 °C (20 °C/min) (2 min). The total run time was 86.0 min. Retention times were established as follow: (*R*)-**8** = 73.8 min, (*S*)-**8** = 75.9 min.

4. Conclusions

SSF cultures were evaluated for kinetic resolution of natural-identical aroma compounds. Agroindustrial side streams proved to be a wholesome growth medium for microorganisms that exhibit high enantioselectivity of hydrolysis of 1-phenylethyl acetate (**1**). Dynamic kinetic resolution of 1-phenylethyl acetate (**1**) catalyzed by *F. oxysporum* AM21 afforded enantiomerically pure (*S*)-**1** (ee > 99%). The (*R*)-enantiomer (ee > 99%) of phenylethanol (**2**) was produced by *A. nidulans* AM243, *P. camemberti* AM83, and *F. avenaceum* AM11. Kinetic resolution of a diastereoisomeric mixture of whisky lactone (**4** and **5**) catalyzed by *A. nidulans* AM243, *F. avenaceum* AM11, and *F. solani* AM203 afforded *trans*-(+)-(4*S*,5*R*)-**4** with ee = 90%. The highest enantioselectivity of hydrolysis of the bicyclic lactone **9** was observed with *P. notatum* AM904 and *B. cirenea* AM235 (ee = 82% and ee = 80% respectively). In conclusion, enantiomerically enriched isomers of aroma lactones can be economically obtained using environmentally friendly techniques by the solid-state fermentation of oilseed cakes.

Acknowledgments: This work was supported by Wroclaw Centre of Biotechnology, programme The Leading National Research Centre (KNOW) for years 2014–2018. We would like to show our gratitude to the students, Malwina Dybiec, Anna Chryplewicz and Bartłomiej Skalny from Students Scientific Association OrgChem, who contributed to the research.

Author Contributions: F.B., E.S., T.O. and A.G. conceived and designed the experiments; E.S. performed the experiments; F.B. and E.S. analyzed the data; F.B. and E.S. wrote the paper.

Conflicts of Interest: The authors declare no conflict of interest.

References

1. Aggelopoulos, T.; Katsieris, K.; Bekatorou, A.; Pandey, A.; Banat, I.M.; Koutinas, A.A. Solid state fermentation of food waste mixtures for single cell protein, aroma volatiles and fat production. *Food Chem.* **2014**, *145*, 710–716. [CrossRef] [PubMed]

2. Kapilan, R. Solid state fermentation for microbial products: A review. *Appl. Sci. Res.* **2015**, *7*, 21–25.

3. Castilho, L.R.; Polato, C.M.S.; Baruque, E.A.; Sant'Anna, G.L.; Freire, D.M.G. Economic analysis of lipase production by *Penicillium restrictum* in solid-state and submerged fermentations. *Biochem. Eng. J.* **2000**, *4*, 239–247. [CrossRef]

4. Joo, H.S.; Kumar, C.G.; Park, G.C.; Paik, S.R.; Chang, C.S. Oxidant and SDS-stable alkaline protease from *Bacillus clausii* I-52: Production and some properties. *J. Appl. Microbiol.* **2003**, *95*, 267–272. [CrossRef] [PubMed]

5. Salihu, A.; Alam, Z.; Abdulkarim, M.I.; Salleh, H.M. Resources, conservation and recycling lipase production: An insight in the utilization of renewable agricultural residues. *Resour. Conserv. Recycl.* **2012**, *58*, 36–44. [CrossRef]

6. De Castro, R.J.S.; Sato, H.H. Enzyme production by solid state fermentation: General aspects and an analysis of the physicochemical characteristics of substrates for agro-industrial wastes valorization. *Waste Biomass Valoriz.* **2015**, *6*, 1085–1093. [CrossRef]

7. Muñoz Solano, D.; Hoyos, P.; Hernáiz, M.J.; Alcántara, A.R.; Sánchez-Montero, J.M. Industrial biotransformations in the synthesis of building blocks leading to enantiopure drugs. *Bioresour. Technol.* **2012**, *115*, 196–207. [CrossRef] [PubMed]

8. Fuhshuku, K.; Oda, S.; Sugai, T. Enzyme reactions as the key step in the synthesis of terpenoids, degraded cartoenoids, steroids, and related substances. *Recent Res. Dev. Org. Chem.* **2002**, *6*, 57–74.

9. Etschmann, M.; Bluemke, W.; Sell, D.; Schrader, J. Biotechnological production of 2-phenylethanol. *Appl. Microbiol. Biotechnol.* **2002**, *59*, 1–8. [CrossRef] [PubMed]

10. Serra, S.; Fuganti, C.; Brenna, E. Biocatalytic preparation of natural flavours and fragrances. *Trends Biotechnol.* **2005**, *23*, 193–198. [CrossRef] [PubMed]

11. Pisani, L.; Superchi, S.; D'Elia, A.; Scafato, P.; Rosini, C. Synthetic approach toward cis-disubstituted γ- and δ-lactones through enantioselective dialkylzinc addition to aldehydes: Application to the synthesis of optically active flavors and fragrances. *Tetrahedron* **2012**, *68*, 5779–5784. [CrossRef]

12. Brenna, E.; Fuganti, C.; Gatti, F.G.; Serra, S. Biocatalytic methods for the synthesis of enantioenriched odor active compounds. *Chem. Rev.* **2011**, *111*, 4036–4072. [CrossRef] [PubMed]

13. Krings, U.; Berger, R.G. Biotechnological production of flavours and fragrances. *Appl. Microbiol. Biotechnol.* **1998**, *49*, 1–8. [CrossRef] [PubMed]

14. Okui, S.; Uchiyama, M.; Mizugaki, M. Metabolism of hydroxy fatty acids II. Intermediates of the oxidative breakdown acid by genus *Candida*. *J. Biochem.* **1963**, *54*, 536–540. [CrossRef] [PubMed]

15. Van der Schaft, P.H.; ter Burg, N.; van den Bosch, S.; Cohen, A.M. Microbial production of natural δ-decalactone and δ-dodecalactone from the corresponding α,β-unsaturated lactones in Massoi bark oil. *Appl. Microbiol. Biotechnol.* **1992**, *36*, 712–716. [CrossRef]

16. Beck, J.J.; Chou, S.C. The structural diversity of phthalides from the *Apiaceae*. *J. Nat. Prod.* **2007**, *70*, 891–900. [CrossRef] [PubMed]

17. Gassenmeier, K.; Schwager, H.; Houben, E.C.R. Unequivocal identification of 1-phenylethyl acetate in clove buds (*Syzygium aromaticum* L.). *Foods* **2017**, *6*, 46. [CrossRef]

18. Alves Macedo, G.; Soberón Lozano, M.M.; Pastore, G.M. Enzymatic synthesis of short chain citronellyl esters by a new lipase from *Rhizopus* sp. *Electron. J. Biotechnol.* **2003**, *6*, 69–72. [CrossRef]

19. Nagy, V.; Toke, E.R.; Keong, L.C.; Szatzker, G.; Ibrahim, D.; Omar, I.C.; Szakács, G.; Poppe, L. Kinetic resolutions with novel, highly enantioselective fungal lipases produced by solid state fermentation. *J. Mol. Catal. B Enzym.* **2006**, *39*, 141–148. [CrossRef]

20. Kamble, M.P.; Chaudhari, S.A.; Singhal, R.S.; Yadav, G.D. Synergism of microwave irradiation and enzyme catalysis in kinetic resolution of (*R,S*)-1-phenylethanol by cutinase from novel isolate *Fusarium* ICT SAC1. *Biochem. Eng. J.* **2017**, *117*, 121–128. [CrossRef]

21. Amin, M.; Bhatti, H.N. Effect of physicochemical parameters on lipase production by *Penicillium fellutanum* using canola seed oil cake as substrate. *Int. J. Agric. Biol.* **2014**, *16*, 118–124.

22. Freitas, A.C.; Castro, R.J.S.; Fontenele, M.A.; Egito, A.S.; Farinas, C.S.; Pinto, G.A.S. Canola cake as a potential substrate for proteolytic enzymes production by a selected strain of *Aspergillus oryzae*: Selection of process conditions and product characterization. *ISRN Microbiol.* **2013**, *2013*, 1–8. [CrossRef] [PubMed]

23. De Castro, A.M.; De Andréa, T.V.; Carvalho, D.F.; Teixeira, M.M. P.; Dos Reis Castilho, L.; Freire, D.M.G. Valorization of residual agroindustrial cakes by fungal production of multienzyme complexes and their use in cold hydrolysis of raw starch. *Waste Biomass Valoriz.* **2011**, *2*, 291–302. [CrossRef]

24. Rehman, S. Optimization of process parameters for enhanced production of lipase by *Penicillium notatum* using agricultural wastes. *Afr. J. Biotechnol.* **2011**, *10*, 19580–19589. [CrossRef]

25. Tan, T.; Zhang, M.; Xu, J.; Zhang, J. Optimization of culture conditions and properties of lipase from *Penicillium camembertii* Thom PG-3. *Process Biochem.* **2004**, *39*, 1495–1502. [CrossRef]

26. Ramachandran, S.; Singh, S.K.; Larroche, C.; Soccol, C.R.; Pandey, A. Oil cakes and their biotechnological applications—A review. *Bioresour. Technol.* **2007**, *98*, 2000–2009. [CrossRef] [PubMed]

27. Mueller, K.; Eisner, P.; Yoshie-Stark, Y.; Nakada, R.; Kirchhoff, E. Functional properties and chemical composition of fractionated brown and yellow linseed meal (*Linum usitatissimum* L.). *J. Food Eng.* **2010**, *98*, 453–460. [CrossRef]

28. Sivaramakrishnan, S.; Gangadharan, D. Edible Oil Cakes. In *Biotechnology for Agro-Industrial Residues Utilisation*; Singh nee' Nigam, P., Pandey, A., Eds.; Springer: New Delhi, India, 2009; pp. 253–271, ISBN 978-1-4020-9942-7.

29. Orsavova, J.; Misurcova, L.; Ambrozova, J.V.; Vicha, R.; Mlcek, J. Fatty acids composition of vegetable oils and its contribution to dietary energy intake and dependence of cardiovascular mortality on dietary intake of fatty acids. *Int. J. Mol. Sci.* **2015**, *16*, 12871–12890. [CrossRef] [PubMed]

30. Bayrak, A.; Kiralan, M.; Ipek, A.; Arslan, N.; Cosge, B.; Khawar, K.M. Fatty acid compositions of linseed (*Linum usitatissimum* L.) genotypes of different origin cultivated in Turkey. *Biotechnol. Biotechnol. Equip.* **2010**, *24*, 1836–1842. [CrossRef]

31. Chen, H.; He, Q. Value-added bioconversion of biomass by solid-state fermentation. *J. Chem. Technol. Biotechnol.* **2012**. [CrossRef]

32. Kumar, S.S.; Gupta, R. An extracellular lipase from *Trichosporon asahii* MSR 54: Medium optimization and enantioselective deacetylation of phenyl ethyl acetate. *Process Biochem.* **2008**, *43*, 1054–1060. [CrossRef]

33. Jadhav, D.D.; Patil, H.S.; Chaya, P.S.; Thulasiram, H.V. Fungal mediated kinetic resolution of racemic acetates to (*R*)-alcohols using *Fusarium proliferatum*. *Tetrahedron Lett.* **2016**, *57*, 4563–4567. [CrossRef]

34. Solarte, C.; Yara-Varón, E.; Eras, J.; Torres, M.; Balcells, M.; Canela-Garayoa, R. Lipase activity and enantioselectivity of whole cells from a wild-type *Aspergillius flavus* strain. *J. Mol. Catal. B Enzym.* **2014**, *100*, 78–83. [CrossRef]

35. Fan, Y.; Xie, Z.; Zhang, H.; Qian, J. Kinetic resolution of both 1-phenylethanol enantiomers produced by hydrolysis of 1-phenylethyl acetate with *Candida antarctica* lipase B in different solvent systems. *Kinet. Catal.* **2011**, *52*, 686–690. [CrossRef]

36. Liang, J.; Zhang, Y.; Sun, A.; Deng, D.; Hu, Y. Enantioselective resolution of (±)-1-phenylethanol and (±)-1-phenylethyl acetate by a novel esterase from *Bacillus* sp. SCSIO 15121. *Appl. Biochem. Biotechnol.* **2016**, *178*, 558–575. [CrossRef] [PubMed]

37. Vandenberghe, A.; Markó, I.E.; Lucaccioni, F.; Lutts, S. Enantioselective hydrolysis of racemic 1-phenylethyl acetate by an enzymatic system from fresh vegetables. *Ind. Crops Prod.* **2013**, *42*, 380–385. [CrossRef]

38. Deng, D.; Zhang, Y.; Sun, A.; Hu, Y. Enantio-selective preparation of (*S*)-1-phenylethanol by a novel marine GDSL lipase MT6 with reverse stereo-selectivity. *Chin. J. Catal.* **2016**, *37*, 1966–1974. [CrossRef]

39. Boratyński, F.; Smuga, M.; Wawrzeńczyk, C. Lactones 42. Stereoselective enzymatic/microbial synthesis of optically active isomers of whisky lactone. *Food Chem.* **2013**, *141*, 419–427. [CrossRef] [PubMed]

40. Boratyński, F.; Dancewicz, K.; Paprocka, M.; Gabryš, B.; Wawrzeńczyk, C. Chemo-enzymatic synthesis of optically active γ- and δ-decalactones and their effect on aphid probing, feeding and settling behavior. *PLoS ONE* **2016**, *11*, e0146160. [CrossRef] [PubMed]

41. Boratyński, F.; Pannek, J.; Walczak, P.; Janik-Polanowicz, A.; Huszcza, E.; Szczepańska, E.; Martinez-Rojas, E.; Olejniczak, T. Microbial alcohol dehydrogenase screening for enantiopure lactone synthesis: Down-stream process from microtiter plate to bench bioreactor. *Process Biochem.* **2014**, *49*. [CrossRef]

42. Walczak, P.; Pannek, J.; Boratyński, F.; Janik-Polanowicz, A.; Olejniczak, T. Synthesis and fungistatic activity of bicyclic lactones and lactams against *Botrytis cinerea*, *Penicillium citrinum*, and *Aspergillus glaucus*. *J. Agric. Food Chem.* **2014**, *62*, 8571–8578. [CrossRef] [PubMed]

catalysts

MDPI

Article

Immobilized *Burkholderia cepacia* Lipase on pH-Responsive Pullulan Derivatives with Improved Enantioselectivity in Chiral Resolution

Li Xu †, Guli Cui †, Caixia Ke, Yanli Fan and Yunjun Yan *

Key Laboratory of Molecular Biophysics, the Ministry of Education, College of Life Science and Technology, Huazhong University of Science and Technology, Wuhan 430074, China; xuli@hust.edu.cn (L.X.); gulicui@hust.edu.cn (G.C.); kecaixia@hust.edu.cn (C.K.); D201577434@hust.edu.cn (Y.F.)
* Correspondence: yanyunjun@mail.hust.edu.cn; Tel.: +86-27-8779-2213
† These authors contributed equally to this work.

Received: 12 December 2017; Accepted: 5 January 2018; Published: 9 January 2018

Abstract: A kind of pH-responsive particle was synthesized using modified pullulan polysaccharide. The synthesized particle possessed a series of merits, such as good dispersity, chemical stability and variability of particle size, making it a suitable carrier for enzyme immobilization. Then, *Burkholderia cepacia* lipase (BCL), a promising biocatalyst in transesterification reaction, was immobilized on the synthesized particle. The highest catalytic activity and immobilization efficiency were achieved at pH 6.5 because the particle size was obviously enlarged and correspondingly the adsorption surface for BCL was significantly increased. The immobilization enzyme loading was further optimized, and the derivative lipase was applied in chiral resolution. Under the optimal reaction conditions, the immobilized BCL showed a very good performance and significantly shortened the reaction equilibrium time from 30 h of the free lipase to 2 h with a conversion rate of 50.0% and ee_s at 99.2%. The immobilized lipase also exhibited good operational stability; after being used for 10 cycles, it still retained over 80% of its original activity. Moreover, it could keep more than 80% activity after storage for 20 days at room temperature in a dry environment. In addition, to learn the potential mechanism, the morphology of the particles and the immobilized lipase were both characterized with a scanning electron microscope and confocal laser scanning microscopy. It was found that the enlarged spherical surface of the particle in low pH values probably led to high immobilized efficiency, resulting in the improvement of enantioselectivity activity in chiral resolution.

Keywords: immobilization; *Burkholderia cepacia* lipase; chiral resolution; 1-phenylethanol; pullulan

1. Introduction

Lipases (triacylglycerol acylhydrolases, EC 3.1.1.3) are a class of promising biocatalysts owing to their catalytic diversity such as hydrolysis, esterification and transesterification [1,2]. Among them, *Burkholderia cepacia* lipase (BCL) is a versatile one and has been widely used in organic synthesis, biodiesel preparation, biodegradation and many other reactions in aqueous and non-aqueous phases [3–5]. Recently, enzymatic chiral resolution by lipases has generated huge interest due to the merits of it having fewer byproducts, a simple operation and environmental friendliness [6,7]. However, the freeform lipases generally exhibit low stability, poor recyclability and non-reusability [3,8], especially low enzyme activity in non-aqueous phase reaction. These demerits were also reflected in the freeform BCL in our previous studies [3,9]. As we know, to employ a proper immobilization method can effectively improve enzyme activity and its dispersibility in solutions. For example, Lee et al. reported that sol–gel immobilized *Candida rugosa* lipase exhibited 10 times higher activity in esterification reaction [10]. So far, a number of materials have been used as immobilization carriers, such as insoluble gel, macroporous resin, carbon nanotubes, and so on. The basic principle is that the physical and chemical properties of the support materials should be stable and harmless to the enzyme. Generally, materials with good dispersibility, hydrophobicity and large surface area are the ideal ones for lipase immobilization [11,12].

Pullulan is a polysaccharide produced by the fungus *Aureobasidium pullulans*. It mainly consists of α-(1-6)-linked maltotriose repeating unit and is easily commercially available [13]. Owing to the advantages of high water solubility, film forming, plasticity and easy biodegradation, pullulan has been widely applied in food and pharmaceutical industries [14]. There is much research on anti-cancer drug delivery using modified pullulan as drug carrier. Many pullulan derivatives have been synthesized for drug delivery systems such as pH/thermo-responsive pullulan-grafted copolymers [15], pullulan nanogel [16] and pullulan-doxorubicin conjugates [17]. It can be inferred that pullulan derivatives are safe, uniformly dispersed and chemically stable, making it a promising immobilization carrier. Furthermore, the pH-responsive pullulan derivatives can significantly increase the immobilization efficiency and the immobilized enzymes can keep relatively high activity owing to the "pH memory" mechanism [18], even under the conditions that they are converted from the aqueous phase into the organic phase. However, since the patent "Enzyme Immobilized with Pullulan Gel" was published in 1981 [19], there have been few reports about the application of pullulan derivatives in enzyme immobilization. Thus, it is necessary and worthy to further explore such kinds of carrier, especially to focus on its sensitivity to pH value.

Moreover, 1-phenylethanol has been utilized in toiletries, chemical industries and many other fields as essential intermediate block. There is much research reporting enzymatic chiral resolution of (*R*, *S*)-1-phenylethanol and the catalytic efficiency can be easily compared between different enzymes [20]. Thus, chiral resolution of (*R*, *S*)-1-phenylethanol has been regarded as one of the model resolution reactions. The enantioselectivity (ee_s), conversion and the duration to achieve reaction equilibrium are the principal factors in evaluating catalytic efficiency. For example, Chen et al. used *Yarrowia lipolytica* lipase to resolve (*R*, *S*)-1-phenylethanol via solvent engineering, increasing the ee_s from 66% to 99.1% in pure n-hexane [21]. In addition, Antonia et al. obtained good conversion rates and high enantioselectivities (92%–>99%) of *Candida antarctica* lipase B using the essentially anhydrous protic ionic liquids in 2012 [22].

Therefore, (*R*, *S*)-1-phenylethanol chiral resolution reaction was utilized to assess the enzyme activity and catalytic efficiency of the freeform and the immobilized lipases. This study attempts to investigate a strategy for BCL immobilization using pullulan derivatives as carrier, to optimize the main affecting parameters on the immobilization efficiency and resolution ability during lipase immobilization, and to further discuss the probable mechanism for pH sensitivity of the immobilized enzyme. Thus, in this study, we aimed to develop a kind of immobilized lipase with controllable particle size by changing the pH value of the solution, which could benefit future application of the immobilized lipase or the combination with other immobilization strategies.

2. Results and Discussion

2.1. Particle Characterization and pH Responsibility

The morphology of the synthesized carrier is shown in Figure 1. The modified pullulan polysaccharide was turned into spherical particles through diafiltration, achieving a good dispersion in reaction mixture. In the view of high magnification, the spherical particles were not a typical smooth sphere but a rough surface. This structure led to a larger specific surface area at 4.57 m$^2 \cdot$g^{-1} measured by BET method in N$_2$ adsorption, which would contribute to higher immobilization efficiency.

As shown in Figure 2, the average particle sizes in different pH values were measured via DLS analysis, indicating the synthesized carriers had very good pH sensitivity. When the solution pH decreased from 10.0 to 4.0, particle sizes were gradually increased, especially within pH values from 7.5 to 6.0, the particle sizes were sharply increased. This phenomenon was mainly due to the introduction of sulfadimethoxine (SDM), a member of the sulfonamide family, which is an important factor responsible for the polymer pH sensitivity [23]. In an acidic environment, SDM was deionized and became hydrophobic, which led to a strong self-assembling of the pullulan derivatives. On the contrary, in the alkaline environment the ionized SDM was hydrophilic, so the pullulan derivatives were rather evenly dispersed than assembled. Therefore, larger particle size was obtained at lower pH while smaller size at higher pH as shown in Figure 2. Moreover, lipase immobilization was achieved within the self-assembling of the synthesized carriers, a larger particle size would lead to a higher immobilization efficiency owing to the occurrence of embedding and adsorption.

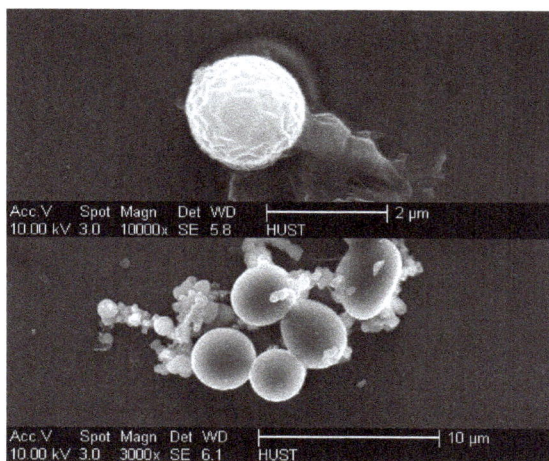

Figure 1. SEM images of the synthesized particles.

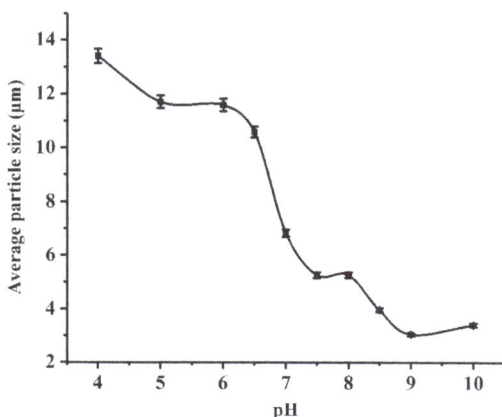

Figure 2. Average particle sizes variation of the synthesized matrix in different solution pH values.

2.2. Lipase Immobilization and Characterization

The immobilized BCL as a catalyst was achieved via physical absorption on the synthesized matrix. The effect of enzyme amount is presented in Figure 3. The ratios between lipase and matrix were ranged from 0.5 to 4.0 and a highest resolution activity was obtained at 2.0. Immobilization efficiency remained nearly 80% in the ratios from 2.0 to 4.0. To minimize cost of the immobilized lipase preparation, the ratio of 2.0 (lipase: matrix) was chosen for later experiments.

Owing to the pH sensitivity of the carriers, the immobilization efficiency was significantly enhanced by changing the pH value during immobilization, thus, the effect of pH on immobilization was investigated (Figure 4). As described in "Section 2.1", lipase was immobilized along with the particle self-assembling through both embedding and adsorption, so the immobilization efficiency was increased with the increment of particle sizes and the highest immobilization efficiency was achieved at pH 4.0. Immobilization pH was not only related to the immobilization efficiency but also to the lipase activity. As the 'pH memory' theory [18] described, when enzymes were immobilized from an aqueous phase, they maintained the ionization, so they could preserve the catalytic activity after immobilization. Meanwhile, it can be seen from Figure 4 that the activity reached its maximum at pH 6.5 but reduced obviously under an acidic environment, even the immobilization efficiency was higher. Thus, the immobilized lipase as catalyst in resolution reaction should be prepared in a nearly neutral buffer with pH 6.5, where the particle size changed greatly.

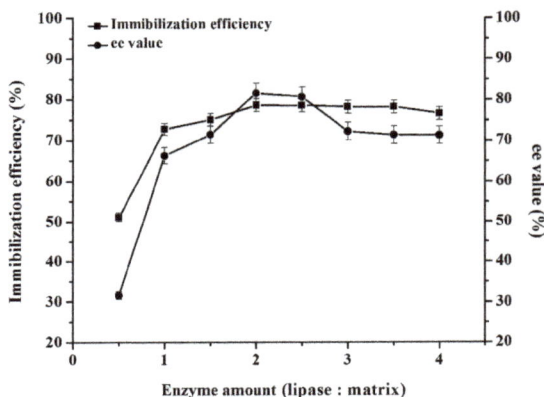

Figure 3. Effects of enzyme amount (lipase: matrix) on the immobilization efficiency and *ee* value in the solution pH of 7. (Reaction conditions: 0.02 g immobilized lipase as catalyst, 45 °C, 1 h).

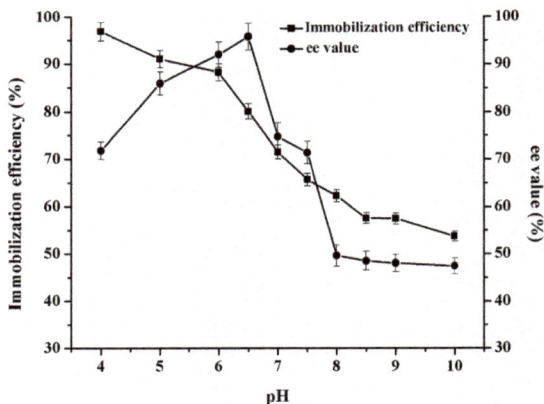

Figure 4. Effects of solution pH values (4–10) on the immobilization efficiency and *ee* value with enzyme amount ratio of 2.0. (Reaction conditions: 0.02 g immobilized lipase as catalyst, 45 °C, 1 h).

In addition, the immobilized BCL was visualized by a confocal laser scanning microscopy (CLSM) [24]. As can be seen from Figure 5, the synthesized carrier displayed as grey component in the vision, and FITC (fluorescein iso-thiocyanate) labeled BCL delivered a green signal under the microscopy, providing a direct evidence for successful immobilization of the lipase.

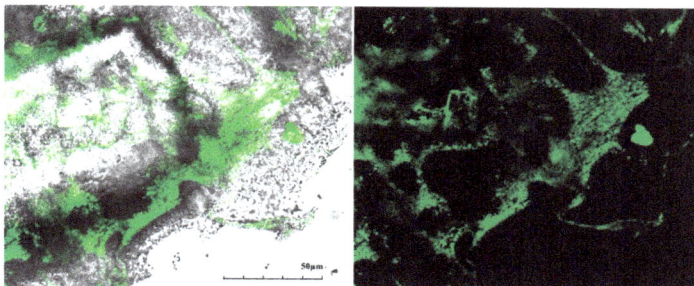

Figure 5. Confocal laser scanning microscopy (CLSM) images of the immobilized BCL labeled with FITC (showing a green signal).

2.3. Catalytic Performance

2.3.1. Effect of Reaction Time

The effect of reaction time on conversion and enantioselectivity (ee_s) in the chiral resolution reaction of (*R*, *S*)-1-phenylethanol is presented in Figure 6. During the catalytic resolution, the conversion and ee_s increased with the extension of reaction time and showed a consistent tendency. At the first 60 min, the reaction had an extremely high reaction rate with the initial one at 79.41 μmol/min·g, and at 90 min the conversion rate and ee_s respectively attained nearly 50% and 90%. After 180 min, the conversion and ee_s were roughly stabilized at a relatively high level, meaning the reaction reached its equilibrium. The conversion rate was nearly 50% and ee_s was close to 100%, indicating that BCL had high enantioselectivity for (*R*)-1-phenylethanol. Nevertheless, the free BCL exhibited very low activity, and only the Pullulan carrier as control didn't show any catalytic activity in this reaction. Additionally, our previous study [9] also proved that the immobilized BCL showed a significant higher initial reaction rate than the freeform.

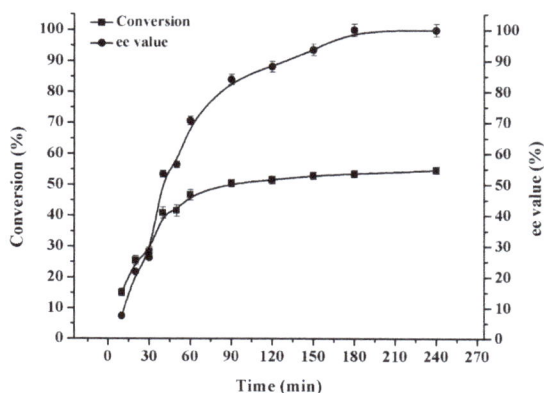

Figure 6. Time curve of resolution (*R*, *S*)-1-phenylethanol using 0.02 g immobilized BCL as catalyst in 45 °C.

2.3.2. Effect of Temperature

The influence of reaction temperature was important to enzyme activity as a proper temperature may lead to the structure change of lipase active site, resulted in enzymatic activation. As shown in Figure 7, the immobilized BCL exhibited a relatively high catalytic activity when the temperature reached 35 °C and then gradually increased. Below 35 °C, lipase protein structure might keep an inactive form owing to the close of the active site. With the increase of environment temperature, the structure was turned into an activated form and the substrates could easily get access to the active site, resulting in a greatly improved catalytic activity. The highest conversion (and ee_s) was achieved at 55 °C, and then the activity decreased beyond this temperature, which probably was due to the inactivation of protein in elevated temperature. In the range of 35–65 °C, the catalytic efficiency kept relatively high where conversion rate was over 45% and ee_s was more than 80%, indicating that the immobilized BCL had a good thermo-stability. It is in good agreement with our previous work [3,9,25].

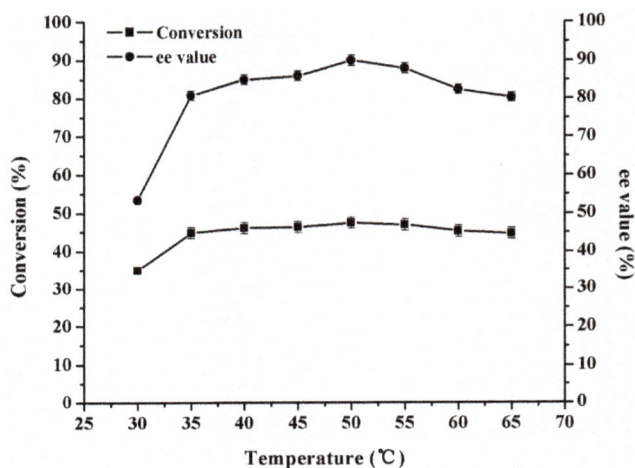

Figure 7. Temperature curve of resolution (*R*, *S*)-1-phenylethanol using 0.02 g immobilized BCL as catalyst for 1 h.

2.3.3. Catalytic Reusability and Storage Stability

It is important for a catalyst to be reusable in industrial application as it can significantly reduce production cost. Thus, catalytic reusability of the immobilized BCL was further explored and is presented in Figure 8. It can be seen that centrifugation could easily separate the catalyst from the reaction system and was reused for 10 cycles. The activity slightly reduced in the first several cycles and then remained above 80% of its original activity, exhibiting very good operationability and reusability. Moreover, the immobilized BCL also presented fairly good storage stability. It could remain over 80% activity after storage for 20 days at room temperature in a drying environment (Figure 9). The above results demonstrated that the immobilized BCL is not only well applicable in research experiments but also possesses a promising prospect in large-scale industrial applications.

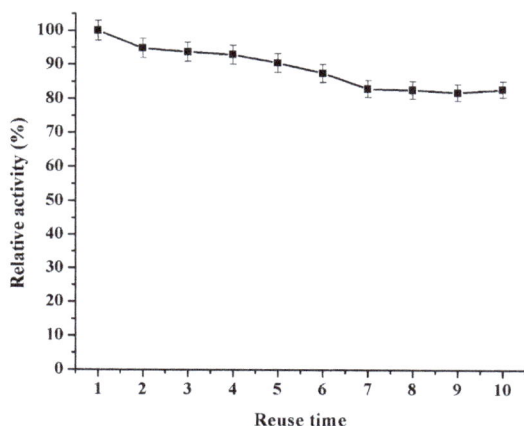

Figure 8. Reusability of immobilized lipase in resolution reaction. (Reaction conditions: 0.02 g immobilized BCL, 45 °C, 1 h).

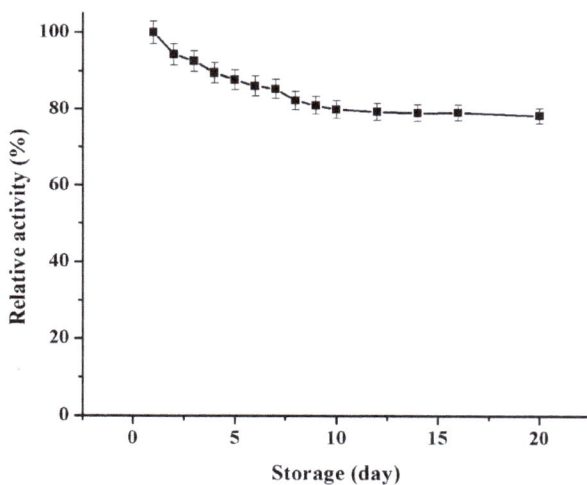

Figure 9. Storage stability of immobilized lipase in resolution reaction. (Reaction conditions: 0.02 g immobilized BCL, 45 °C, 1 h).

2.4. In Comparison with Other Enzymes

Pullulan has been widely employed in immunity therapy researches as a promising carrier of protein drug. However, there were few reports on immobilization of enzyme with pullulan since the patent "Enzyme Immobilized with Pullulan Gel" was published in 1981. In recent years, Swati and Rekha synthesized a kind of alginate beads made from pullulan-complexed α-amylase and glucosidase via entrapment immobilization method [26]. The addition of pullulan obviously enhanced the immobilization efficiency and the beads of enzymes–pullulan complex showed higher enzyme activity in hydrolysis of starch than the corresponding beads of free enzymes. In this work, the resolution efficiency of BCL was largely improved by immobilizing it on the modified pullulan particles. Under the optimal reaction conditions, resolution equilibrium could be achieved within 2 h where conversion rate reached 50% and the corresponding ee_s was 99.2%. While in the same conditions, the free BCL needs 30 h to reach the reaction equilibrium.

It can be seen from Table 1, compared with related works on chiral resolution published in the past decades [9,27–32], the immobilized BCL in this work exhibited a higher enantioselectivity and a shorter reaction time (2 h). As BCL was immobilized through physical absorption, the enhancement of resolution efficiency was mainly attributable to the properties of pH sensitive particles. During the lipase immobilization, a proper pH value could lead to a sharply increase in particle size, which would provide even larger spherical surface for lipase immobilization, resulting in higher immobilization efficiency as well as easier access of substrates to the lipase. Moreover, the synthesized particles could be uniformly dispersed in the reaction system, and easily separated by centrifugation for the subsequent reuse. These results suggest that the modified pullulan is a suitable immobilization carrier and has a good potential for lipase immobilization in the future.

Table 1. Comparison of the enantioselectivity of other enzymes.

Lipase from	Immobilization Carrier	Conversion (%)	ee_s (%)	Time (h)	References
C. rugosa	Magnetic β-cyclodextrin nanoparticles	49.5	98.0	24	28
C. rugosa	Magnetic chitosan nanoparticles	50.8	79.1	140	29
Candida sp. 99–125	Aminopropyl-grafted mesoporous silica nanotubes	NM *	80.0	90	30
B. cepacia	Hybrid with calcium phosphate	47.8	91.1	24	27
B. cepacia	Active carbon cloth	46.0	32.0	24	31
B. cepacia	pH-responsive particles	50.0	99.2	2	This work
B. cepacia	Free lipase	50.0	99.0	30	9
P. cepacia	Free lipase	7.1	7.7	48	32

* NM: not measured in the report.

3. Experimental

3.1. Materials

Burkholderia cepacia lipase (BCL) powder was bought from Amano, Nagoya-shi, Aichi, Japan. Pullulan polysaccharide was obtained from Aladdin, Shanghai, China. (*R*, *S*)-1-phenylethanol was bought from Sigma Aldrich Co., Ltd., St. Louis, MO, USA. High-performance liquid chromatography (HPLC) grade organic solvents were purchased from TEDIA, Fairfield, OH, USA. Acetic anhydride, succinic anhydride, vinyl acetate, n-heptane and other reagents were bought from Sinopharm Chemical Reagent Co., Ltd., Shanghai, China.

3.2. Synthesizing pH-Responsive Particles from Pullulan Polysaccharide

As reported by Kun and You [33], pullulan polysaccharide was firstly modified by acetylation forming pullulan acetate (PA). Then, PA was succinylated and conjugated with sulfadimethoxine (SDM). Finally, the polymers were self-assembled into hydrogel particles using a diafiltration method, and then the pullulan derivative was obtained. Subsequently, the particles were filtered and dried in a thermostatic vacuum drier. The synthesized carrier turned into a kind of white precipitation and then it was ground into powder for later use.

3.3. Particle Size and Specific Surface Area Measurement

Dynamic light scattering (DLS) analysis was utilized to measure the average particle size of the resultant carrier. Before the DLS examination, particle samples were incubated in solution buffer with different pH for 12 h. To investigate the effect of pH on particle size, three different kinds of solution buffers were used. They were Na_2HPO_4–citrate acid buffer (0.2 M, pH 4.0–5.0), sodium phosphate buffer (0.2 M, pH 6.0–8.0), and Tris–HCl buffer (0.2 M, pH 9.0–10.0). In addition, the specific surface areas of the synthesized particles were also performed and calculated using BET (Brunauer–Emmett–Teller) method in N_2 adsorption [34].

3.4. Lipase Immobilization

The synthesized particles were used as the immobilization carrier. A proper quantity of lipase was dissolved in 5 mL buffer solution containing 0.1 g immobilization carrier in a 50 mL tube. The mixture was incubated with rotary shaking under 37 °C at 7.2 g for 12 h. The supernatant were removed by centrifugation at 4 °C, 25,800 g for 10 min. The frozen vacuum drier was used to dehydrate the immobilized lipase and then it was ground into powder for later use. During this procedure, the effect of lipase amount on immobilization efficiency was examined and optimized. The protein content of the supernatant was mensurated via the Bradford's method using bovine serum albumin (BSA) as standard [35]. Immobilization efficiency (%) was estimated via Equation (1).

$$\text{Immobilization efficiency (\%)} = \frac{\text{immobilized protein}}{\text{total loading protein}} \times 100\% \tag{1}$$

3.5. Catalyzing Resolution of (R, S)-1-Phenylethanol and Efficiency Evaluation

1 mmol racemic 1-phenylethanol, 4 mmol vinyl acetate and 5 mL pure heptane were mixed together to start the reactions (Scheme 1). The mixture and 0.02 g immobilized or free BCL as catalysts were added into a 50 mL stoppered flask, shaking at 45 °C, 7.2 g for 60 min. Meanwhile, only the Pullulan carrier without the enzyme was also used in this reaction as a control. During the catalyzing process, the effects of reaction time, temperature and pH on resolution efficiency and the performances of reuse and storage were totally examined. After the reactions, the catalysts were removed by centrifugation at 4 °C, 25,800 g for 2 min. A 0.45 μm filter was used to filter the samples and HPLC was used to analyze all the samples. The resolution reaction mentioned above was utilized to measure the enzyme activity. One unit (U) of enzyme activity was defined as the amount of enzyme that produces 1 μmol α-phenylethyl acetate in one minute under the assay conditions. According to our previous work [36], substrate and product were analyzed by HPLC (Model 2300-525, Scientific Systems, Inc. (SSI), State College, PA, USA) with a Chiral OD-H column (4.6 mm × 250 mm, Daicel Chemical, Tokyo, Japan). The mobile phase was composed of hexane/2-propanol alcohol at 95/5 (*v*/*v*) and the flow rate was 1.0 mL·min^{-1}. The samples were detected at 254 nm (Model 525 UV Detector, Scientific Systems, Inc. (SSI), State College, PA, USA). In the above condition, (*R*)- and (*S*)-1-phenylethanol were separated and appeared in different resulted peaks of HPLC between 7 and 10 min. The same conditions were used to run all the samples as stated above.

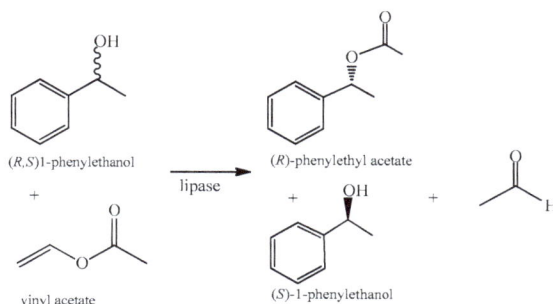

Scheme 1. Lipase catalyzed resolution of (R, S)-1-phenylethanol.

As Chen et al. [37] described, enantioselectivity was represented by E value and it was calculated by Equation (2); substrate enantiomeric excess (ee_s) was estimated by Equation (3), and substrate conversion (C) computed by Equation (4).

$$E = \frac{\ln[(1-C)(1-ee_s)]}{\ln[(1-C)(1+ee_s)]} \tag{2}$$

$$ee_s = \frac{S-R}{S+R} \tag{3}$$

$$C = \frac{S_0 + R_0 - (S+R)}{S_0 + R_0} \tag{4}$$

where, S_0 and R_0 respectively represent the concentrations of the (S)- and (R)- enantiomers of 1-phenylethanol before the reaction; S and R represent the concentrations of the (S)- and (R)-enantiomers of 1-phenylethanol after the reaction.

3.6. Characterization via SEM and CLSM for the Probable Mechanism for pH Sensitivity

Morphology characterization of the synthesized particles was analyzed by scanning electron microscope (Nova Nano SEM 450, FEI Company, and Eindhoven, The Netherlands). The samples were coated with gold using a sputter coating system and tested under an acceleration voltage of 5 kV.

The immobilized lipase on the synthesized particles was examined through a confocal laser scanning microscopy (CLSM) with a SIM scanner (Olympus FV1000 Co., Tokyo, Japan) to test the fluorescencent signal from the FITC (fluorescein iso-thiocyanate)-marked immobilized BCL. The labeling procedure was as follows [38]: Firstly, lipase was dissolved in sodium phosphate buffer solution (pH 7.0, 0.05 M) with a final concentration at 10 mg·mL^{-1} and fluorescein iso-thiocyanate diffused in dimethyl sulfoxide (DMSO) and the final concentration was 1 mg·mL^{-1}. Secondly, the mixture of lipase solution (5 mL) together with FITC (100 μL) was incubated at 4 °C without light for 24 h. Lastly, the residual FITC was separated from the mixture by dialyzing with double distilled water and the immobilized lipase labeled with FITC was achieved.

4. Conclusions

In this study, a new particle was synthesized using diafiltration method from the modified pullulan polysaccharide. The size of the derivative particle exhibited pH sensitivity and increased along with the decrease of solution pH, rendering itself a series of unique merits in enzyme immobilization. BCL, a promising biocatalyst in transesterification reaction, was immobilized on the synthesized particle and employed in (R, S)-1-phenylethanol chiral resolution. The highest enantioselectivity was achieved at pH 6.5 because the particle size was obviously increased, leading to a larger specific surface area and resulting in easier access of substrates to active site of the immobilized lipase. The immobilized BCL also exhibited very good operation stability; after 10 reuse cycles, about 80% original activity still remained, and over 80% relative activity remained after storage for 20 days at room temperature in a drying environment. More importantly, it could sharply shorten the reaction time to 2 h with a conversion rate of 50.0% and ee_s at 99.2%, much better than the ones reported in the literature. This study indicates that the immobilized BCL exhibits good catalysis performance and possesses a prosperous prospect in future industrial application.

Acknowledgments: This work is financially supported by the National Natural Science Foundation of China (grant Nos. 31070089, 31170078 and J1103514), the National High Technology Research and Development Program of China (grant Nos. 2013AA065805 and 2014AA093510), the National Natural Science Foundation of Hubei Province (grant No. 2015CFA085), and the Fundamental Research Funds for HUST (grant Nos. 2014NY007, 2017KFYXJJ212, 2017KFXKJC010, 2017KFTSZZ001). Many thanks are indebted to the Analytical and Testing Center of Huazhong University of Science and Technology for their valuable assistances in the FTIR, TEM, and XRD measurements.

Author Contributions: Li Xu and Yunjun Yan conceived and designed the experiments; Caixia Ke and Guli Cui performed the experiments; Guli Cui analyzed the data and wrote the paper; Yanli Fan assisted in part of the experiments; Guli Cui, Li Xu and Yunjun Yan contributed to the revision and proofreading of the manuscript.

Conflicts of Interest: The authors declare no conflict of interest.

References

1. Singh, A.K.; Mukhopadhyay, M. Overview of Fungal Lipase: A Review. *Appl. Biochem. Biotechnol.* **2012**, *166*, 486–520. [CrossRef] [PubMed]

2. Jaeger, K.E.; Eggert, T. Lipases for biotechnology. *Curr. Opin. Biotechnol.* **2002**, *13*, 390–397. [CrossRef]

3. Liu, T.; Liu, Y.; Wang, X.; Li, Q.; Wang, J.; Yan, Y. Improving catalytic performance of *Burkholderia cepacia* lipase immobilized on macroporous resin NKA. *J. Mol. Catal. B Enzym.* **2011**, *71*, 45–50. [CrossRef]

4. Pan, S.; Xue, L.; Xie, Y.; Yi, Y.; Chong, L.; Yan, Y.; Yun, L. Esterification activity and conformation studies of *Burkholderia cepacia* lipase in conventional organic solvents, ionic liquids and their co-solvent mixture media. *Bioresour. Technol.* **2010**, *101*, 9822–9824. [CrossRef] [PubMed]

5. Yang, J.; Guo, D.; Yan, Y. Cloning, expression and characterization of a novel thermal stable and short-chain alcohol tolerant lipase from *Burkholderia cepacia* strain G63. *J. Mol. Catal. B Enzym.* **2007**, *45*, 91–96. [CrossRef]

6. Drauz, K.; Waldmann, H. *Enzyme Catalysis in Organic Synthesis: A Comprehensive Handbook*; Wiley-VCH: Hoboken, NJ, USA, 2002; pp. 991–1033.

7. Ghanem, A.; Aboul-Enein, H.Y. Lipase-Mediated Chiral Resolution of Racemates in Organic Solvents. *Cheminform* **2004**, *15*, 3331–3351. [CrossRef]

8. Wang, J.; Ma, C.; Bao, Y.; Xu, P. Lipase entrapment in protamine-induced bio-zirconia particles: Characterization and application to the resolution of (*R*, *S*)-1-phenylethanol. *Enzyme Microb. Technol.* **2012**, *51*, 40–46. [CrossRef] [PubMed]

9. Li, X.; Huang, S.; Xu, L.; Yan, Y. Improving activity and enantioselectivity of lipase via immobilization on macroporous resin for resolution of racemic 1-phenylethanol in non-aqueous medium. *BMC Biotechnol.* **2013**, *13*, 92. [CrossRef] [PubMed]

10. Sang, H.L.; Doan, T.; Won, K.; Ha, S.H.; Koo, Y.M. Immobilization of lipase within carbon nanotube–silica composites for non-aqueous reaction systems. *J. Mol. Catal. B Enzym.* **2010**, *62*, 169–172. [CrossRef]

11. Jesionowski, T.; Zdarta, J.; Krajewska, B. Enzyme immobilization by adsorption: A review. *Adsorption* **2014**, *20*, 801–821. [CrossRef]

12. Homaei, A.A.; Sariri, R.; Vianello, F.; Stevanato, R. Enzyme immobilization: An update. *J. Chem. Biol.* **2013**, *6*, 185–205. [CrossRef] [PubMed]

13. Prajapati, V.D.; Jani, G.K.; Khanda, S.M. Pullulan: An exopolysaccharide and its various applications. *Carbohydr. Polym.* **2013**, *95*, 540–549. [CrossRef] [PubMed]

14. Singh, R.S.; Kaur, N.; Kennedy, J.F. Pullulan and pullulan derivatives as promising biomolecules for drug and gene targeting. *Carbohydr. Polym.* **2015**, *123*, 190–207. [CrossRef] [PubMed]

15. Fundueanu, G.; Constantin, M.; Ascenzi, P. Preparation and characterization of pH- and temperature-sensitive pullulan microspheres for controlled release of drugs. *Biomaterials* **2008**, *29*, 2767–2775. [CrossRef] [PubMed]

16. Hirakura, T.; Yasugi, K.; Nemoto, T.; Sato, M.; Shimoboji, T.; Aso, Y.; Morimoto, N.; Akiyoshi, K. Hybrid hyaluronan hydrogel encapsulating nanogel as a protein nanocarrier: New system for sustained delivery of protein with a chaperone-like function. *J. Control. Release* **2010**, *142*, 483–489. [CrossRef] [PubMed]

17. Li, H.; Bian, S.; Huang, Y.; Liang, J.; Fan, Y.; Zhang, X. High drug loading pH-sensitive pullulan-DOX conjugate nanoparticles for hepatic targeting. *J. Biomed. Mater. Res. A* **2014**, *102*, 150–159. [CrossRef] [PubMed]

18. Costantino, H.R.; Griebenow, K.; Langer, R.; Klibanov, A.M. On the pH memory of lyophilized compounds containing protein functional groups. *Biotechnol. Bioeng.* **1997**, *53*, 345–348. [CrossRef]

19. Hirohara, H.; Nabeshima, S.; Fujimoto, M.; Nagase, T. Enzyme Immobilization with Pullulan Gel. U.S. Patent 4247642, 27 January 1981.

20. Chua, L.S.; Sarmidi, M.R. Immobilised lipase-catalysed resolution of (*R*, *S*)-1-phenylethanol in recirculated packed bed reactor. *J. Mol. Catal. B Enzym.* **2004**, *28*, 111–119. [CrossRef]

21. Cui, C.; Xie, R.; Tao, Y.; Zeng, Q.; Chen, B. Improving performance of *Yarrowia lipolytica* lipase lip2-catalyzed kinetic resolution of (*R, S*)-1-phenylethanol by solvent engineering. *Biocatal. Biotransform.* **2015**, *33*, 38–43. [CrossRef]

22. Ríos, A.P.D.L.; Rantwijk, F.V.; Sheldon, R.A. Effective resolution of 1-phenyl ethanol by *Candida antarctica* lipase B catalysed acylation with vinyl acetate in protic ionic liquids (PILs). *Green Chem.* **2012**, *14*, 1584–1588. [CrossRef]

23. Park, S.Y.; Bae, Y.H. Novel pH-sensitive polymers containing sulfonamide groups. *Macromol. Rapid Commun.* **1999**, *20*, 269–273. [CrossRef]

24. Paddock, S.W. Confocal laser scanning microscopy. *Biotechniques* **1999**, *27*, 992–1004. [PubMed]

25. Ke, C.; Li, X.; Huang, S.; Xu, L.; Yan, Y. Enhancing enzyme activity and enantioselectivity of *Burkholderia cepacia* lipase via immobilization on modified multi-walled carbon nanotubes. *RSC Adv.* **2014**, *4*, 57810–57818. [CrossRef]

26. Jadhav, S.B.; Singhal, R.S. Pullulan-complexed α-amylase and glucosidase in alginate beads: Enhanced entrapment and stability. *Carbohydr. Polym.* **2014**, *105*, 49–56. [CrossRef] [PubMed]

27. Ke, C.; Fan, Y.; Chen, Y.; Xu, L.; Yan, Y. A new lipase-inorganic hybrid nanoflower with enhanced enzyme activity. *RSC Adv.* **2016**, *6*, 19413–19416. [CrossRef]

28. Ozyilmaz, E.; Sayin, S.; Arslan, M.; Yilmaz, M. Improving catalytic hydrolysis reaction efficiency of sol–gel-encapsulated *Candida rugosa* lipase with magnetic β-cyclodextrin nanoparticles. *Colloids Surf. B* **2014**, 182–189. [CrossRef] [PubMed]

29. Siódmiak, T.; Ziegler-Borowska, M.; Marszałł, M.P. Lipase-immobilized magnetic chitosan nanoparticles for kinetic resolution of (*R, S*)-ibuprofen. *J. Mol. Catal. B Enzym.* **2013**, *94*, 7–14. [CrossRef]

30. Bai, W.; Yang, Y.; Tao, X.; Chen, J.; Tan, T. Immobilization of lipase on aminopropyl-grafted mesoporous silica nanotubes for the resolution of (*R, S*)-1-phenylethanol. *J. Mol. Catal. B Enzym.* **2012**, *76*, 82–88. [CrossRef]

31. Hara, P.; Mikkola, J.; Murzin, D.Y.; Kanerva, L.T. Supported ionic liquids in *Burkholderia cepacia* lipase-catalyzed asymmetric acylation. *J. Mol. Catal. B Enzym.* **2010**, *67*, 129–134. [CrossRef]

32. Xue, P.; Yan, X.H.; Wang, Z. Lipase immobilized on HOOC-MCF: A highly enantioselective catalyst for transesterification resolution of (*R, S*)-1-phenylethanol. *Chin. Chem. Lett.* **2007**, *18*, 929–932. [CrossRef]

33. Na, K.; You, H.B. Self-assembled hydrogel nanoparticles responsive to tumor extracellular pH from pullulan derivative/sulfonamide conjugate: Characterization, aggregation, and adriamycin release in vitro. *Pharm. Res.* **2002**, *19*, 681–688. [CrossRef] [PubMed]

34. Park, E.Y.; Sato, M.; Kojima, S. Fatty acid methyl ester production using lipase-immobilizing silica particles with different particle sizes and different specific surface areas. *Enzyme Microb. Technol.* **2006**, *39*, 889–896. [CrossRef]

35. Kruger, N.J. The Bradford method for protein quantitation. In *Methods in Molecular Biology*; Walker, J.M., Ed.; Humana Press: London, UK, 1994; Volume 32, pp. 9–15.

36. Li, X.; Xu, L.; Wang, G.; Zhang, H.; Yan, Y. Conformation studies on *Burkholderia cenocepacia* lipase via resolution of racemic 1-phenylethanol in non-aqueous medium and its process optimization. *Process Biochem.* **2013**, *48*, 1905–1913. [CrossRef]

37. Chen, C.S.; Fujimoto, Y.; Girdaukas, G.; Sih, C.J. Quantitative analyses of biochemical kinetic resolutions of enantiomers. *J. Am. Chem. Soc.* **1982**, *104*, 7294–7299. [CrossRef]

38. Feng, S.; Li, G.; Fan, Y.; Yan, Y. Enhanced performance of lipase via microcapsulation and its application in biodiesel preparation. *Sci. Rep.* **2016**, *6*, 29670. [CrossRef]

catalysts

MDPI

Review

Recent Advances in ω-Transaminase-Mediated Biocatalysis for the Enantioselective Synthesis of Chiral Amines

Mahesh D. Patil [1], Gideon Grogan [2], Andreas Bommarius [3] and Hyungdon Yun [1,*]

[1] Department of Systems Biotechnology, Konkuk University, 120 Neungdong-ro, Gwangjin-gu, Seoul 05029, Korea; mahi1709@gmail.com

[2] Department of Chemistry, University of York, Heslington, York YO10 5DD, UK; gideon.grogan@york.ac.uk

[3] School of Chemical & Biomolecular Engineering, Georgia Institute of Technology, 311 Ferst Drive, Atlanta, GA 30332-0100, USA; andreas.bommarius@chbe.gatech.edu

* Correspondence: hyungdon@konkuk.ac.kr

Received: 24 May 2018; Accepted: 19 June 2018; Published: 21 June 2018

check for updates

Abstract: Chiral amines are important components of 40–45% of small molecule pharmaceuticals and many other industrially important fine chemicals and agrochemicals. Recent advances in synthetic applications of ω-transaminases for the production of chiral amines are reviewed herein. Although a new pool of potential ω-transaminases is being continuously screened and characterized from various microbial strains, their industrial application is limited by factors such as disfavored reaction equilibrium, poor substrate scope, and product inhibition. We present a closer look at recent developments in overcoming these challenges by various reaction engineering approaches. Furthermore, protein engineering techniques, which play a crucial role in improving the substrate scope of these biocatalysts and their operational stability, are also presented. Last, the incorporation of ω-transaminases in multi-enzymatic cascades, which significantly improves their synthetic applicability in the synthesis of complex chemical compounds, is detailed. This analysis of recent advances shows that ω-transaminases will continue to provide an efficient alternative to conventional catalysis for the synthesis of enantiomerically pure amines.

Keywords: biocatalysis; chiral amines; ω-transaminases; protein engineering; reaction engineering; multi-enzymatic cascades

1. Introduction

Enantiomerically pure amines are indispensable constituents of various small molecule pharmaceuticals, fine chemicals, and agrochemicals [1]. The increasing demands for enantiopure compounds in the pharmaceutical, fine chemicals, and agricultural industries, together with the environmental restrictions approved by many economies, require the effective integration of traditional chemical synthesis methods with those of enzymatic *'greener'* approaches [2–5]. The chemical synthesis of chiral amines is labor intensive, requiring harsh reaction conditions and use of toxic intermediates. The use of toxic metals also warrants a purification step, thus further complicating the synthetic systems and the cost thereof [6]. As enzymes are obtained from renewable resources, they fulfil the basic demands of sustainable and green chemistry, proposed by Graedel et al. [7]. In this context, biocatalysis, the application of enzymes for efficient and selective chemical transformations, has been recognized as a major *'green technology'* that provides sustainable synthetic methods towards a range of chiral compounds [8,9].

Transaminases (TAs) are one of the largest groups of enzymes used in the synthesis of chiral amine building blocks [10–13]. As TAs play key roles in various cellular signaling and metabolic pathways, their presence is ubiquitous in nature [14]. The history of transamination can be traced back to 1930, when Needham and coworkers observed the relationship between levels of amino acids such as L-glutamic acid, L-aspartic acid and oxaloacetic acid in the breast muscles of pigeon [15]. Since the first demonstration of transamination in the late 1930s, various transaminases have been discovered in subsequent decades [16,17]. Depending on their specificities towards various substrates, transaminases are generally classified as α- or ω-TAs. Transaminases are classified based on the structures of the amino donor substrates, which are divided based on the presence and position of a negatively charged group with respect to the amino group being transferred. In the case of α-Transaminases (i.e., amino acid TAs), the presence of a carboxylic acid group in the α-position to the keto or amine functionality is exclusively required [18]. ω-TAs transfer an amino group attached to a primary carbon at least one carbon away from a carboxyl group (Figure 1). In addition, compounds lacking any carboxyl groups, such as amines and ketones, can serve as substrates for ω-TAs, thus making these enzymes potential biocatalysts for the synthesis of chiral amine compounds [18–20]. ω-TAs accepting amines/ketones without carboxylic group are interchangeably referred to as amino transaminases.

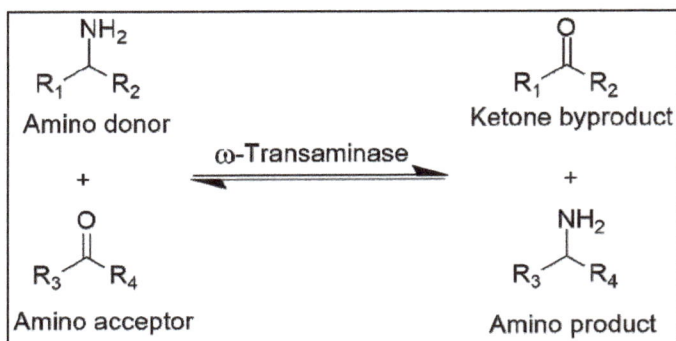

Figure 1. Schematic of transamination catalyzed by transaminases.

The mechanism of the ω-TA-catalyzed transamination reaction is well studied and has been reported to exhibit ping-pong-bi-bi kinetics. In a two-step reaction, Pyridoxal 5′-phosphate (PLP), bound to the lysine residue of an enzyme, mediates the transfer of amino group from an amino donor, such as L-alanine to form pyridoxamine phosphate (PMP) intermediate. Subsequently, this amine group is transferred to the amino acceptor, such as a keto acid or ketone and the PLP is regenerated for the next catalytic cycle [12,18,21]. The enantiomeric purity of the stereospecific isomer formed by TA-mediated biocatalysis is a consequence of their exclusive substrate-binding mode, which is constrained by the relative position of PLP and two substrate-binding pockets of different sizes.

PLP-dependent enzymes are divided into seven Fold-Types (I–VII) and despite their wide substrate specificities, ω-transaminases are only found in two different protein folds: (S)-selective enzymes, which belong to protein fold-type I and (R)-selective enzymes belonging to protein fold-type IV [19,20,22].

2. Synthetic Approaches for Chiral Amines and Equilibrium Shift for ω-TA-Mediated Biocatalysis

ω-TAs can be employed for the synthesis of enantiopure amines using three approaches: (A) kinetic resolution of the racemic mixture of amines; (B) asymmetric synthesis using pro-chiral ketones and (C) deracemization of racemic amines (Figure 2). The previous two decades have seen a surge of interest in TA-mediated amination reactions.

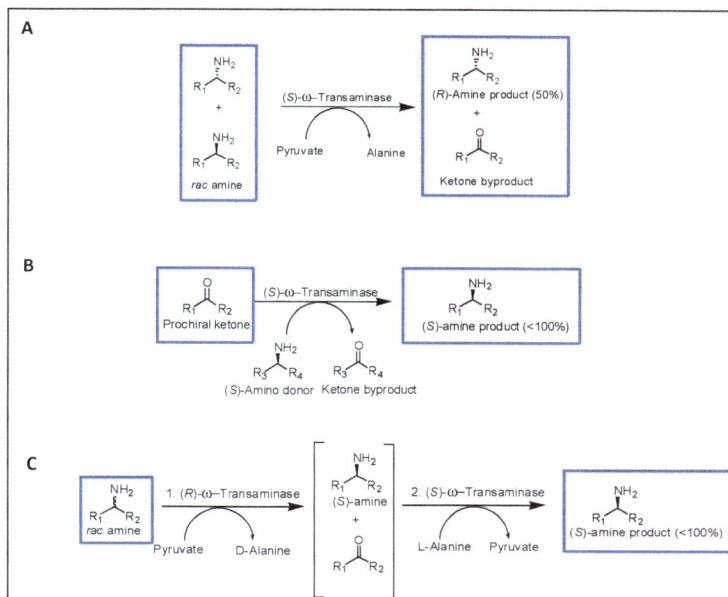

Figure 2. Synthetic approaches for ω-TA-mediated amine synthesis: (**A**) Kinetic resolution with (*S*)-ω-TA; (**B**) Asymmetric synthesis and (**C**) Deracemization of racemic amine using (*R*)-ω-TA and (*S*)-ω-TA. (Adapted from [14], Copyright (2010), with permission from Elsevier).

2.1. Kinetic Resolution

The kinetic resolution of primary amines employs stoichiometric amounts of an amino acceptor, wherein the reaction equilibrium favors product formation. Since the early 1990s, when Celgene Corporation successfully used kinetic resolution for the synthesis of enantiopure amines [23], this approach has been extensively explored in the last few years [15,24–28].

With the developing substrate scope and improved enzymatic characteristics, many *"ready to use"* screening toolkits are commercially available for their application in biocatalysis. For instance, commercially available ω-TAs from Codexis such as ATA-256, ATA-114, ATA-117, ATA-113, and ATA-103 have been successfully applied for the synthesis of chiral amines [12,29,30]. In addition, a new pool of potential TAs is being continuously screened and characterized from various microbial strains [31,32]. Recently, an (*S*)-ω-transaminase from the thermophilic eubacterium *Sphaerobacter thermophilus* was expressed and functionally characterized in Yun's laboratory [33]. This thermostable enzyme was used for the kinetic resolution of various amines, and β- and γ-amino acids, in the presence of pyruvate as amino acceptor. Wu et al. have characterized four new distinct ω-TAs from *Pseudomonas putida* NBRC 14164, which were utilized for the kinetic resolution of racemic amines such as 4-phenylbutan-2-amine, 2,3-dihydro-1*H*-inden-1-amine and amino alcohols such as 2-amino-2-phenylethanol, 2-aminobutan-1-ol, etc. [24]. More recently, Iglesias et al. [34] characterized an (*R*)-ω-TA from *Caproniasemi immersa*. In their studies, whole cell extracts of *E. coli* BL21 (DE3) expressing (*R*)-ω-TA were utilized for the resolution of several racemic amines in the presence of pyruvic acid as an amine acceptor. Most recently, Baud et al. used a metagenomics approach for the discovery of new ω-TAs and subsequently for the synthesis of allylic amines [35]. Three enzymes were selected among the 11 putative class-III TAs identified by genome mining. These enzymes were further evaluated for their transamination potential of functionalized cinnamaldehydes, which are ultimately used as building blocks for antifungal Naftin® (naftifine) and analogs thereof [35]. Gao et al. also identified and characterized a novel

(R)-enantioselective ω-TA from *Fusarium oxysporum* by genome mining and evaluated it for the kinetic resolution of amine substrates, such as 1-(3-chlorophenyl)ethanamine, 1-(2-fluorophenyl)ethanamine and (4-chlorophenyl)(phenyl)methanamine [36].

Although kinetic resolution is advantageous in requiring only a single enzyme, the atom efficiency of kinetic resolution remains low with a maximum theoretical yield of 50% [15]. The kinetic resolution also suffers other disadvantages, such as the formed ketone-product from corresponding amines being reported to cause severe inhibition [14]. Thus, the removal of formed ketone to circumvent product inhibition is highly desirable (Figure 3(AI)). Various strategies, such as physical extraction/vaporization, are available to remove the ketone products and thus to shift the equilibrium of the reversible reaction towards product formation [37] (Figure 3). Accordingly, isopropylamine has been identified as an industrial choice, since it is readily available and economically viable, and its product acetone is volatile and can easily be removed by reduced pressure (Figure 3(AII)) [34,38,39]. However, complete removal of low concentrations of acetone (<100 mM), without removing other volatile components of the reaction, is a challenging task.

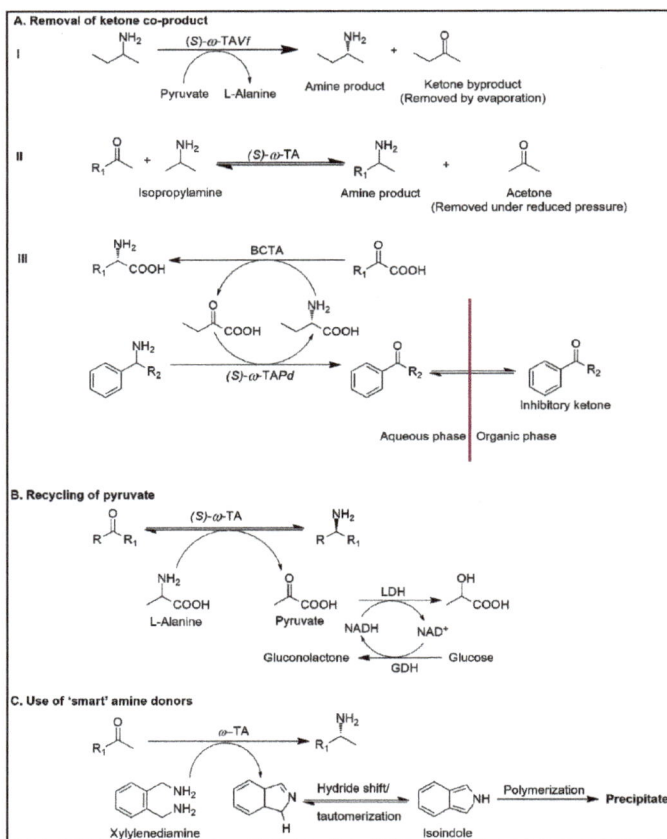

Figure 3. Various strategies used to shift the equilibrium of ω-TA mediated transamination: (**AI**) Removal of ketone byproduct by evaporation; (**AII**) Removal of ketone byproduct under reduced pressure; (**AIII**) Removal of ketone byproduct using biphasic reaction system; (**B**) Recycling of pyruvate substrate; and (**C**) Use of *'smart'* amine donors (BCTA- Branched-chain amino acid transaminases; LDH-Lactate dehydrogenase; GDH-Glucose dehydrogenase).

It has also been reported that the removal of ketones by physical methods is critically affected by the equilibrium constant and volatility of the ketones. Owing to the differential polarities of amines and ketones, methods such as extraction using bi-phasic reactions systems, either liquid-solid or liquid-liquid, have been developed for the continuous removal of ketone products (Figure 3(AIII)) [40]. However, extraction using organic/aqueous biphasic systems is limited by the denaturation of enzymes in an organic environment and the poor discrimination between substrate and product [12]. Nonetheless, to avoid enzyme denaturation in biphasic systems, stabilization of enzymes by immobilization [41,42] (detailed in Section 4.3), enzyme membrane- and packed bed reactors have also been reported [15,43,44].

2.2. Asymmetric Synthesis

The maximum theoretical yield of 50% limits the practical applications of the kinetic resolution approach. Thus, despite being more challenging than kinetic resolution, asymmetric synthesis and deracemization are preferred strategies, as they can generate a theoretical yield of 100% [45]. In asymmetric synthesis, prochiral ketones are aminated to the corresponding chiral amines, with the possibility of obtaining the desired enantiopure amine form in 100% yield [14].

Other than an industrially applied procedure for antidiabetic Sitagliptin, the recent successful examples of amine synthesis by ω-TA-mediated catalysis include pyridyl alkylamines [46], (*R*)-ramatroban precursor [47] and serinol [48]. Moreover, substituted aminotetralins, potential agents used to treat Parkinson's disease and cardiovascular disorders, have also been successfully synthesized using an engineered ω-TA from *Arthrobacter citreus* [ω-TA*Ac*] [49] and a novel (*S*)-selective ω-TA from *Pseudomonas fluorescens* KNK08-18 [50]. However, ω-TA-mediated transamination reactions generally are thermodynamically limited [51]. For instance, the reported equilibrium constant of ω-TA from *Vibrio fluvialis* JS17 (ω-TA*Vf*) for the kinetic resolution of (*S*)-α-MBA, using pyruvate as co-substrate, is 1135. Thus, the equilibrium is greatly shifted toward the formation of L-alanine and acetophenone, consequently making it difficult to perform asymmetric synthesis of (*S*)-α-MBA without rapid product removal and/or the use of sacrificial amines as donors [52].

Subsequently, various strategies have been applied to prevail over these limitations. For instance, a 20-fold excess of (*S*)-α-MBA was used in the asymmetric synthesis of serinol-monoester, a precursor for the synthesis of chloramphenicol, wherein the (*S*)-enantiomer was obtained with 92% conversion and 92% ee, when F85L/V153A variant of ω-TA*Vf* was used [48]. In addition, many of the ω-TAs use L-alanine as an amine donor. Thus, the recycling or removal of pyruvate formed can help to overcome the thermodynamic barrier of the reaction. The application of two strategies is generally used for the removal of pyruvate co-product. (1) Use of lactose dehydrogenase (LDH) and NADH recycling system [53], wherein pyruvate is reduced to lactate, which is recycled using glucose dehydrogenase at the expense of cheaper substrate i.e., glucose (Figure 3B). (2) Use of alanine dehydrogenase (AlaDH) for the effective recycling of L-alanine as amine donor has been reported to reduce the cost of amine donor by 97%, while simultaneously decreasing the E-factor (environmental factor; the ratio of the mass of waste per mass of product) and improving the atom efficiency [54]. However, the use of additional enzymes and their cofactors leads to further complications of the overall biocatalytic reactions.

Although α-MBA, isopropylamine and L-alanine have been most frequently used as amine donors in the ω-TA-mediated asymmetric syntheses, more reactive amine donors are highly desirable to drive reaction equilibrium, and to reduce the use of a high molar excess of these sacrificial amine donors used in the biocatalytic reactions. Recently, '*smart*' diamine donors, such as xylylenediamine (XDA) were discovered, which is readily converted to an amino aldehyde. The subsequent intramolecular cyclization of the amino aldehydes tautomerizes to form isoindol, which is polymerized, and can easily be removed from the reaction mixture due to precipitation (Figure 3C) [53]. However, XDA is not accepted by all wild-type ω-TAs, and it is not a cheaper amino donor, ultimately increasing the cost of the biocatalytic reaction [30]. It was more recently reported that only 1.5 equiv. of *cis*-but-2-ene-1,4-diamine can be used as sacrificial amine donor for the conversion of aromatic sec-alcohols with different substitutions in the aromatic ring to the corresponding amines in a sequential application of laccases and ω-TAs [55]. Furthermore, recyclable biogenic terminal diamines, such as putrescine, cadaverine, and spermidine have also recently been discovered to drive the equilibrium towards amination of ketones. Following the transamination, these bio-based diamines are converted into reactive amino aldehydes, which spontaneously get converted to cyclic imines, thus shifting the equilibrium towards product formation [56].

The synthesis of chiral amines using ω-TA-mediated catalysis has been explored using broad substrate types, including linear, cyclic and aromatic ketones. Numerous research groups have exploited this biocatalytic route for the synthesis of a wide range of pharmaceutical compounds, their intermediates and unnatural amino acids such as beta-and gamma-amino acids [33,57]. Various methods, for instances, co-product removal using cascade enzymes, and the use of 'smart' amine donors to reduce the excess use of sacrificial amines, have been developed to overcome the limitations in biocatalytic reactions employing ω-TAs. Recently, Rehn et al. demonstrated the successful use of supported liquid membrane for the in-situ product removal of (S)-α-methylbenzylamine (MBA) produced by immobilized ω-TAAc [58]. Other strategies, such as substrate engineering [59] and protein engineering, have also been developed for the efficient synthesis of chiral amines by ω-TA-mediated biocatalysis.

2.3. Deracemization

While asymmetric synthesis is still the most favored approach to biocatalytic amine synthesis, deracemization is also attractive, especially when access to racemic amines as substrates is easier than to the corresponding prochiral ketone. In a deracemization reaction, theoretical complete conversion and enantioselectivity is achieved by employing two stereocomplementary ω-TAs. The first step in the deracemization process affords the formation of $\leq$50% enantiopure amine and the corresponding ketone co-product by enantioselective deamination of the racemic amine. This is followed by the asymmetric amination of the ketone by an enantiocomplementary ω-TA, forming the optically pure amine in up to 100% yield [17]. Deracemization has been effectively used for the synthesis of important pharmaceutical drugs, such as mexiletine, an orally effective anti-arrhythmic drug (Figure 4) [60,61]. In this one-pot two-step process on 100 mg scale, an isolated yield of 97% with excellent enantioselectivity (>99% ee) could be achieved by the combination of two stereocomplementary enzymes [61]. In addition, the change in an absolute configuration of the generated product could easily be achieved and both (R)- and (S)-enantiomers can be produced by this method, notably without compromising the isolated yield or enantioselectivity, by switching the order of transaminases applied.

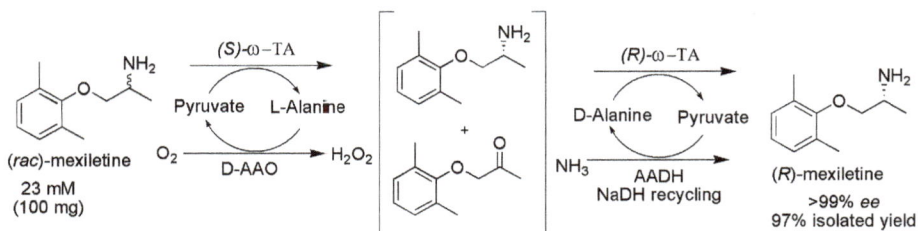

Figure 4. Use of a deracemization strategy for the synthesis of enantiopure Mexiletine (reprinted with permission from [61], Copyright (2009) American Chemical Society). (Reaction conditions: 50 mM racemic substrate, 50 mM sodium pyruvate, 20 mg whole cells of *E. coli* expressing ω-TAs, phosphate buffer (100 mM, pH 7.0), 1 mM PLP, 30 °C, 24 h).

The application of ω-TAs in dynamic kinetic resolutions has also been reported to be very effective for the synthesis of production of Niraparib, a potential anti-cancer drug, and the smoothened receptor inhibitor (SMO), used in the treatment of leukemia [62,63]. Deracemization of various important amines such as 4-phenylbutan-2-amine, a precursor for the antihypertensive dilevalol; sec-butylamine, 1-methoxy-2-propanamine and 1-cyclohexylethylamine, precursors of inhibitors of tumour necrosis factor-α (TNF-α); and 1-phenyl-1-propylamine, a precursor of anti-depressant agent corticotropin releasing factor type-1 receptor antagonist, and other important aromatic amines has been reported in the literature [64,65].

3. Protein Engineering Aspects for Improved Substrate Scope of ω-Transaminase (ω-TA) Biocatalysts

In the context of ω-TAs, the scientific community has used various protein engineering aspects for the efficient synthesis of chiral amines, to improve the substrate scope of available enzymes and further even to alter the stereopreference of the enzymes [66]. In this section, we will discuss the recent advances in the protein engineering aspects of transaminase-mediated biocatalysis for the synthesis of chiral amines drugs and drug intermediates.

The knowledge obtained from the structure of various ω-TAs has established that, generally, the active form of ω-TA is a homodimer, and the active site is positioned at the interface of the monomers, wherein each monomer participates in the active site [67]. The substrate binding region of ω-TAs is composed of two pockets: a large pocket that allows accommodation of substituents with rather broad size distribution, such as small alkyl to naphthyl groups and small binding pocket allows access to the small substituents, not exceeding that of a methyl group [48,68]. Although this characteristic architecture of the active site limits the substrate scope, it also contributes to the high stereopreference of ω-TAs [69]. Until very recently, the number of (*R*)-selective ω-TAs was very limited and much of the effort has been directed towards protein engineering approaches for the generation of stereospecific ω-TAs, and to broaden their substrate scope. The first successful industrially applicable protein engineering approach was the development of the (*R*)-selective ω-TA-117 from *Arthrobacter* sp. for the synthesis of the antidiabetic drug Sitagliptin [29]. The substrate ketone, pro-sitagliptin, is not a natural substrate and could not be accepted by wild-type enzyme. Extensive computational modeling, along with a directed evolution approach, was utilized to incorporate 27 mutations, not only in the active site of the enzyme but also in the interface of the enzyme dimer. This active variant was used at the pilot-scale production of Sitagliptin, wherein 92% yield with excellent enantioselectivity (>99.95% *ee*) was afforded by 6 gL^{-1} enzyme load. This improved process resulted in the reduction of total waste produced by 19%, which was acknowledged by its achievement of the '*Presidential Green Chemistry Challenge Award for Greener Reaction Conditions*' in 2010 [70].

Since the successful engineering of ω-TA-117, many research groups have directed their scientific efforts to the engineering of active site pockets for the acceptance of bulkier substrates by TAs. For instance, the work by Midelfort et al. was directed towards making TAs amenable biocatalysts for synthesis of β-aminoesters from the corresponding β-ketoesters. The (*S*)-ω-TA*Vf* was engineered for the synthesis of (3*S*,5*R*)-ethyl 3-amino-5-methyloctanoate; a key intermediate in the synthesis of imagabalin, a drug candidate in Phase III clinical trials for the treatment of generalized anxiety disorder. The final improved variant, r414, contained eight mutations (F19W/W57F/F85A/R88K/V153A/K163F/I259V/R415F), which displayed a 60-fold increase in initial reaction velocity for the transamination of (*R*)-ethyl 5-methyl 3-oxooctanoate to (3*S*,5*R*)-ethyl 3-amino-5-methyloctanoate (Figure 5) [71].

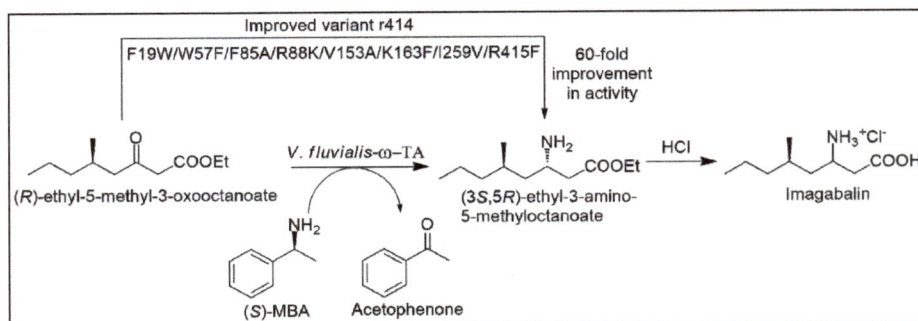

Figure 5. Improved production of Imagablin by protein engineering of (*S*)-ω-TA*Vf* [71]. (Reaction conditions: 4.0 g of wet *E. coli* cells containing (*S*)-ω-TA*Vf* mutant r414, 88 mL potassium phosphate buffer (0.1 M, pH 7.0); 54 mg pyridoxal phosphate; 1 g substrate, 7.5 mL of 1 M solution of (*S*)-α-MBA, 30 °C, 24 h).

Further efforts to improve the substrate scope of (*S*)-ω-TA*Vf* to accommodate the bulky branched-chain substrates were carried out by Bornscheuer et al. [72]. In these studies, the most suitable mutation sites and amino acid substitutions were identified by bioinformatic analysis using 3DM. The volume of the substrate binding pocket was enlarged by the creation of a library that included mainly hydrophobic residues in the active site. The best variant containing four mutations (L56V, W57C, F85V, V153A) was able to catalyze the asymmetric synthesis of sterically demanding branched-chain chiral amines, which afforded complete conversion of the ketone and yielded the corresponding amine with >99% *ee*, notably with a preference for the (*R*)-enantiomer.

In another study, (*S*)-ω-TA*Vf* was engineered using a rational design strategy combining *in silico* and *in vitro* studies for the asymmetric synthesis of (*S*)-1-(1,1′-biphenyl-2-yl)ethanamine by ω-TA-mediated transamination between bulky aromatic ketone substrate 2-acetylbiphenyl and the amine donor isopropylamine [73]. The incorporation of two mutations (W57G/R415A) generated a mutant that showed detectable enzyme activity, and the rationally designed variant (W57F/R88H/V153S/K163F/I259M/R415A/V422A) improved the reaction rate by >1716-fold towards the bulky ketone 2-acetylbiphenyl, producing the corresponding (*S*)-amine with excellent enantiomeric purity (>99% *ee*).

Since the substrate specificity of ω-TAs is largely controlled by steric constraints in the substrate-binding pockets, bulky α-keto acids carrying a side chain bigger than an ethyl substituent generally are not accepted by ω-TAs. The (*S*)-selective ω-TA from *Ochrobactrum anthropic* ((*S*)-ω-TA*Oa*) was engineered by employing alanine scanning mutagenesis to reduce steric constraints and thereby to allow the small pocket to readily accept bulky substituents. Alanine substitution at residue L57 (L57A), a residue located in the large pocket, induces an altered binding orientation of a carboxyl group of keto acids and amino acids and thus was able to provide more room for the small pocket. This variant allowed the acceptance of bulkier substrates, 2-oxopentanoic acid, and L-norvaline and improved the activity towards these substrates by 48-and 56-fold, respectively. The applicability of the generated mutant was evaluated for the kinetic resolution of racemic norvaline and furthermore to produce optically pure L- and D-norvaline by asymmetric amination of 2-oxopentanoic acid [74]. In another study, Nobili et al. carried out systematic mutagenesis study of the active site residues of (*S*)-ω-TA*Vf* to expand its substrate scope towards two bulky ketones. This group identified two active mutants, F85L/V153A and Y150F/V153A, which showed a 30-fold increased activity in the conversion of (*S*)-phenylbutylamine and (*R*)-phenylglycinol, respectively [68]. However, saturation mutagenesis at the respective position in (*S*)-ω-TA*Vf* did not improve the activity towards the same substrates studied in the previous studies, implying that the substrate specificity of ω-TA is a complex episode and cannot be directly related from one enzyme template to another.

In a contribution to expand the substrate scope of ω-TAs, Pavlidis et al. [75] mutated a fold class I (*S*)-ω-TA from *Ruegeria* sp. TM1040 [(*S*)-ω-TA*Ru*; PDB ID: 3FCR] by rational design and identified an active variant with just four mutations (3FCR_QM; Y59W/Y87F/Y152F/T231A), which proved sufficient for the acceptance of bulkier ketones and enabled asymmetric synthesis of a wide set of bulky chiral amines on a preparative scale with excellent conversion, isolated yield and stereoselectivity. This group further carried out mutational studies on (*S*)-ω-TA*Ru* for the synthesis of a bridged bicyclic amine *exo*-3-amino-8-aza-bicyclo[3.2.1]oct-8-yl-phenyl-methanone, a motif constituent of many neuroactive agents, from its corresponding ketone substrate [76]. Two active variants were identified (3FCR_QM/I234F) and (3FCR_QM/I234M), which exhibited specific activity of 80 and 116 mU mg^{-1}, respectively, while maintaining absolute stereoselectivity compared to 45 mU mg^{-1} of the mutant identified in the previous studies (3FCR_QM) for the kinetic resolution of a bicyclic amine *exo*-3-amino-8-aza-bicyclo[3.2.1]oct-8-yl-phenyl-methanone.

Although (*S*)-ω-TA*Ru* has successfully been engineered to accept aromatic (such as bicyclic) and thus planar bulky amines, the variants active towards the substrates carrying a bulky tert-butyl substituent adjacent to the carbonyl function, such as 2,2-dimethyl-1-phenyl-propan-1-one, were unavailable until very recently. To address this issue, (*S*)-ω-TA*Ru* was engineered by introducing five mutations [Y59W/Y87L/T231A/L382M/G429A] [77]. This active variant enabled the asymmetric synthesis of (*R*)-2,2-dimethyl-1-phenylpropan-1-amine from the corresponding ketone substrate 2,2-dimethyl-1-phenyl-propan-1-one and isopropylamine as an amine donor. These studies suggested the critical importance of amino residue Y87, as it blocks the entrance for the sterically demanding tert-butyl moiety, and the introduction of a mutation at this position (Y87L) allowed the hydrophobic interaction between the substrate moiety and amino acid residue at position 87, thus leading to the accommodation of the bulky substrate [77].

Protein engineering approaches have played a crucial role in the generation of highly efficient ω-TAs. Various approaches, such as directed evolution and rationally designed engineering of ω-TAs, not only improved the scope for bulkier substrates but also improved their enantioselectivity. Knowledge of the various putative sequences of protein families has successfully been used for the identification of novel transaminases. Various approaches have been utilized for improvement in kinetic resolution and asymmetric synthesis approaches. The removal of co-products by physical extraction using membrane bioreactors has been successfully implemented to favor the reaction equilibrium towards product formation. Furthermore, reaction engineering by employing cascade reactions to remove and/or recycle the product have also found a useful role in equilibrium shift. To this end, various 'smart' amine donors have been developed that allow the cyclization or the tautomerization. Ever since the successful demonstration of protein engineering for the industrial production of sitagliptin, various approaches of protein engineering have spurred improvement in the substrate scope of TAs and their enantioselectivities.

4. Establishing Stable ω-TA Biocatalysts for Synthetic Applications

The synthetic applicability of TA biocatalysts can be successfully extended to industrial scale only with the use of robust and reusable biocatalysts, which can offer excellent stability in non-physiological conditions. The stability of any enzymatic preparation can be exemplified as its thermodynamic stability (conformational stability of the enzyme structure), kinetic stability (stability over time) and operational stability, which includes thermostability and stability in organic solvents [78,79]. The following sections summarize the various strategies adopted to improve the stability of ω-TAs and thereby their applicability in the synthesis of chiral amines.

4.1. Finding Thermostable ω-TAs

Owing to their excellent stability, thermostable enzymes permit execution of the biocatalytic reaction at higher temperatures. This results in the achievement of higher reaction rates and furthermore to the easier removal of volatile by-products, thereby helping to shift the equilibrium of the reaction towards product formation [11,41]. Enzymes originating from thermophilic microorganisms naturally possess desired properties of high operational and thermodynamic stabilities [80]. Recently, Cerioli and co-workers [81] identified a ω-TA from the halophilic bacterium *Halomonas elongata* exhibiting highest activity at 50 °C. In addition, the enzyme activity was retained in the presence of co-solvents up-to 20%, proving the enzyme suitable for industrial application of transamination of poorly water-soluble substrates. Chen et al. [82] improved the operational stability of ω-TA from *Chromobacterium violaceum* (ω-TACv) by the addition of additives, co-solvents, and organic solvents, and demonstrated the first successful application of ω-TA co-immobilized with surfactants. It was shown that the enzyme retains its active dimeric structure when it is stored in 50% glycerol, 20% methanol or 10% DMSO. Mathew et al. recently identified two thermostable ω-TAs, each from *Thermomicrobium roseum* (ω-TA*Tr*) and *Sphaerobacter therlmophilus*(ω-TA*St*) respectively [33,83]. A (*S*)-ω-TA*Tr* showed improvement in residual activity when the enzyme was incubated for 1 h at 60 and 70 °C, which was attributed to the refolding of an enzyme to its natural conformation at higher temperatures. In addition, this enzyme exhibited a broad substrate scope and could be utilized for the synthesis of chiral amines using asymmetric synthesis and kinetic resolution approaches [83]. Another thermostable TA, (*S*)-ω-TA*St*, showed the highest activity at 60 °C and no loss of activity following 1 h incubation at this temperature was noticed. It was also reported that the stability of (*S*)-ω-TA*S* was not adversely affected in the presence of 20% DMSO, implying the industrial potential of the enzyme [33]. More recently, Chen et al. [84] identified a thermostable ω-TA from *Geobacillus thermodenitrificans* (ω-TA*Gt*), which showed highest activity at 65°C. The ω-TA*Gt* also exhibited 2.5-fold increase in activity when the enzyme was incubated for 8 h at 50 °C or 30 min at 60 °C. Most recently, Ferrandi et al. [85] discovered the novel thermostable ω-TA by exploiting the metagenomics approach. One of the three discovered ω-TAs, B3-TA, was exceptionally thermostable and retained 85% and 40% of the initial activity after the incubation at 80 °C for 5 days and 2 weeks, respectively. B3-TA is the most thermostable natural ω-TA known till date, which also showed excellent tolerance toward different water-miscible and water immiscible organic solvents (Table 1).

Table 1. Selected examples of recently discovered thermostable ω-TAs.

Sr. No.	Source	Enzyme Selectivity	Strategy Used	Optimum Temperature (°C)	Remarks	Reference
1	*Halomonas elongata* DSM 2581	(*S*)-selective	Conserved domain analysis	50	First study reported on ω-TA from the moderate halophile bacterium *H. elongate*. The stability was unaffected in the presence of organic solvents.	[81]
2	*Chromobacterium violaceum*	(*S*)-selective		65	The enzyme performance was improved 5-fold by a co-lyophilization with surfactants	[82]
3	*Sphaerobacter thermophilus*	(*S*)-selective	BLAST search against protein sequences from thermophiles	60	The enzyme was utilized for the stereoselective synthesis of β- and γ- amino acids	[33]
4	*Thermomicrobium roseum*	(*S*)-selective	BLAST search against protein sequences from thermophiles	87	Volatile inhibitory byproducts were removed by performing asymmetric synthesis and kinetic resolution at high temperature	[83]
5	*Geobacillus thermodenitrificans*	(*S*)-selective	BLAST search against protein sequences from thermophiles	65	The enzyme showed relatively good activity toward ketoses, suggesting its potential for catalyzing the asymmetric synthesis of chiral amino alcohols.	[84]
6	Hot spring metagenomes	(*S*)-selective	Metagenomics	88	The most thermostable natural ω-TA known till date.	[85]

4.2. Protein Engineering

In addition to its application to the improvement of the substrate scope of ω-TAs, protein engineering approaches have also been extended to improve the operational stability of the enzymes (Table 2). For instance, (*R*)-selective ω-TA-117 from *Arthrobacter* sp., employed in the industrial production of sitagliptin, was engineered to improve its tolerance towards organic solvents [29]. Eleven rounds of evolution generated the optimized biocatalyst, which could afford a successful transamination of high substrate load (50 g L^{-1} pro-sitagliptin ketone) at 45 °C, importantly, in the presence of 50% DMSO as a co-solvent. Pannuri et al. also engineered ω-TA*Ac* to improve its thermostability and subsequently used it for stereoselective synthesis of chiral amines [86].

The recent structural studies of ω-TAs have established the fact that the homotetrameric conformations of ω-TAs possess superior operational stability compared to their homodimeric counterparts [87,88]. It has also been postulated that the structural integrity of the cofactor-ring binding domain is a decisive factor in the operational stability of ω-TAs. The cofactor-ring motif, positioned at the interface of monomer and dimer interface, is known to shape the bottom of the active site and to shield the cofactor from the solvent environment. To gain deeper insights into the role of cofactor-ring motif in the operational stability of tetrameric ω-TAs, Börner et al. [89] recently generated several mutant libraries by targeting 32 amino acid residues around the active site and cofactor-ring domains. A double mutant (Asn161Ile and Tyr164Leu) increased the resting stability of the enzyme by 11 °C compared to its wild-type parent enzyme. Notably, mutations introduced at these two positions conferred an excellent operational stability in biphasic reaction system at 45 °C. Moreover, the addition of third mutation (Gly51Ser) granted an additive effect in terms of thermal resistance and improved the thermal resistance of the resting state by 4 °C [89].

In another study, the (*R*)-ω-TA from *Aspergillus terreus* [(*R*)-ω-TA*At*] was recently engineered by employing the combination of an *in silico* strategy and mutational studies by site-directed mutagenesis [90]. The potential mutation sites were selected based on the folding energy calculations and four stabilized mutants (T130M, T130F, E133F and D134L) were obtained. These studies were successful in generating a single mutant (T130M) and double mutant (T130M/E133F), which improved the half-life (t$_{1/2}$) at 40 °C by ~2.2 and 3.3-fold, respectively. The mutations carried out at these positions generated new hydrophobic clusters and increased the hydrogen bonds, thereby improving the thermostability of the mutants compared to the wild-type protein.

Protein engineering by incorporation of noncanonical amino acids has also been proven to be an attractive strategy to improve the stability and activity of ω-TAs [91]. Recently, Yun's group successfully used this approach for the improvement of enzymatic properties and thereby its synthetic applications of (*S*)-ω-TA*Vf*. The global incorporation of 3-fluorotyrosine resulted in 2.3-fold improvement of a half-life of engineered enzyme compared to its wild-type counterpart. Furthermore, the engineered (*S*)-ω-TA*Vf* could completely convert 25 mM of acetophenone into (*S*)-MBA with excellent enantioselectivity (*ee* >99%) in the presence of 20% DMSO (*v/v*), which was ~2-fold higher than wild-type ω-TA*Vf* [92].

Table 2. Selected examples of application of protein engineering techniques to improve operational stability of ω-TAs.

Sr. No.	Source	Enzyme Selectivity	Strategy Used	Remarks	Reference
1	*Arthrobacter* sp.	(R)-selective	Substrate walking, modeling, and mutation approach	11 rounds of evolution generated the optimized biocatalyst, which could afford a successful transamination of in the presence of 50% DMSO as a co-solvent	[29]
2	*Arthobacter citreus*	(S)-selective	Structure-guided enzyme mutagenesis	The enzyme was used for the industrial production of substituted aminotetraline	[86]
3	*Novel enzyme from c-LEcta'smetagenomic library*		Semi-rational mutagenesis	32 amino acid residues were targeted around the active site including the cofactor-ring motif for superior operational, thermo- and solvent stability	[89]
4	*Aspergillus terreus*	(R)-selective	*In silico* design employing B-factor and folding free energy calculations	The optimum catalytic temperature of a mutant T130F was increased by 10 °C	[90]
5	*Sphaerobacter thermophilus*	(S)-selective	Residue-specific andsite-specific incorporation of unnatural amino acids	The residue-specific incorporation showed 2-fold enhancement in the half life at 70 °C	[91]
6	*Vibrio fluvialis* JS17	(S)-selective	Global incorporation of unnatural amino acids	Mutant exhibited ~2-fold higher tolerance towards 20% DMSO compared to wild-type	[92]

4.3. Enzyme Immobilization

A single enzyme can differentially perform in different formulations in terms of activity and stability. Although the cell-free, purified form of a protein is the most desirable form of any biocatalysts, the higher purification costs limit its application on large scales. Biocatalyst in another form i.e., cell free extracts, can also be used if the target protein is sufficiently overexpressed. However, the long-term storage stability of the purified as well as cell free extracts is poor [12]. Therefore, use of the whole cells expressing the biocatalyst of interest is of potential use. However, the major limitation of the use of whole cell biocatalysts is the poor accessibility of the substrate to the enzymes located within the cells [93].

One potential approach to improve the stability of biocatalysts is their immobilization. Immobilization is advantageous owing to its low downstream requirements, as it allows the easy recovery of the biocatalysts and their reuse in the subsequent batch or continuous reactions [42,93,94]. Immobilization has been widely used to improve the operational and storage stability of ω-TAs (Table 3). For instance, Koszelewski et al. [95] encapsulated the commercially available enzymes TA-117, TA-113, TA-114 and ω-TAVf using sol–gel/celite matrix for the first time. The immobilized TA-117 exhibited excellent operational stability and retained about 78% of the initial activity after eight repetitive uses, most importantly, without compromising the enantioselectivity. The immobilization by covalent attachment using various support materials such as chitosan is also widely reported [96–99]. De Souza et al. [100] recently utilized cellulose for immobilization of ω-TAVf. The ω-TAVf immobilized in the epoxy-functionalized cellulose conferred excellent temperature stability and reusability, wherein enzyme was consistently active over the broad range of temperature from 30 to 60 °C and for four repetitive cycles in the asymmetric synthesis of (S)-phenylethylamine.

The entrapment of whole cells expressing ω-TAs using natural materials such as Ca-alginate, gelatin, or agarose has been reported. For instance, Rehn et al. compared the various methods of immobilization of recombinant *E. coli* cells expressing ω-TAAc. Chitosan proved to be the method of choice for the whole-cell immobilization as it retained >90% activity after eight successive batches [42]. In another study, Cardenas-Fernandez et al. successfully overcame the diffusional limitations of whole-cell immobilization by entrapping the permeabilized whole cells in PVA-gel (Lentikats®) [101].

While most of the non-commercial supports limit their application because of the labor-intensive preparations prior to immobilization, commercially available ready-to-use supports such as polymeric resins are feasible. Neto et al. [41] recently immobilized (S)-ω-TA (ATA-47) and (R)-ω-TA (Ate-TA) using commercially available polemeric resins as enzyme carriers. Among six, two enzyme carriers i.e., (Relizyme HA403; pore size-40–60 nm and Supabeads EC-HA; pore size-10–20 nm; both having hexamethylamino functional groups) allowed the reuse of the preparation for 250 h of operation with retention of more than 50% of the initial activity. In addition, the immobilized preparations increased thermal stability, allowing storing the enzyme for more than 60 days at room temperature [41]. Most recently, Abaházi et al. [1] covalently immobilized Try60Cys mutant of ω-TACv on bisepoxide-activated polymeric resins. The enzyme mutant immobilized by linking it to polymeric resin with ethylamine function activated with glycerol diglycidyl ether-EA-G could preserve the activity even in the high concentration of DMSO (50% *v/v*). Furthermore, this enzyme preparation could be used for 19 consecutive cycles in batch mode for the kinetic resolution of *rac*-4-phenylbutan-2-amine to its corresponding *R*-enantiomer, a precursor for antihypertensive dilevalol, with excellent activity and selectivity (C~50%, *ee* > 99%).

Protein fusion tags, such as polyhistidine tags, provide an efficient protein purification by affinity chromatography and allow the enzyme immobilization in assembling immobilized enzyme microreactors (IEMR) [11,102,103]. IEMRs are the unique platforms for *in vitro* multi-enzymatic synthetic pathways. Furthermore, recent advances in nanotechnology have provided several nanomaterials for the efficient immobilization of the enzymes. Nanomaterials provide the large surface area that improves the adsorption of proteins and thereby allow the separation of protein in a low-concentration samples [104] Immobilization of ω-TAs using Ni^{2+}-chelated polydopamine magnetic nanoparticles [104], magnetic PVA-Fe_3O_4 nanoparticles [105] and MnO_2 nanorods [106] has been recently reported in the literature.

Table 3. Selected examples of immobilization techniques used to improve the operational stability of ω-TAs.

Sr. No.	Source	Type of Immobilization	Support Used for Immobilization	Comment	Reference
1	*Chromobacterium violaceum* (Trp60Cys mutant)	Covalent binding	bisepoxide-activated polymeric resins	Immobilized preparation was used for 19 consecutive reaction cycles and in media containing up to 50% (v/v) DMSO as co-solvent in batch mode reactions.	[1]
2	*(S)-selective ATA-47*	Ionic interaction	Relizyme HA403 (commercial material)	Immobilized preparation retained more than 50% initial activity for 8 cycles (each of 24 h; corresponding to more than 250 h of operation)	[41]
3	*(R)-selective Atc-TA*	Hydrophobic	Supabeads EC-HA (commercial material)	The immobilized preparation increased the thermal stability, allowing storing the enzyme for more than 60 days	[41]
4	*ω-TA 117*	Sol-gel entrapment	Celite 545	The immobilized enzyme preparation could be recycledeight times with only moderate decrease of activity for each cycle	[95]
5	*Gibberella zeae*	Covalent binding	Modified chitosan	Significant improvement in pH and temperature stability	[96]
6	*Neosartorya fischeri*	Covalent binding	Modified chitosan	Significant improvement in pH and temperature stability	[96]
7	*Aspergillus fumigatus*	Covalent binding	Chitosan	The immobilized enzyme showed higher activity at 70 °C compared to free enzyme	[97]
8	*Ruegeria pomeroyi*	Covalent binding	Chitosan	The immobilized enzyme retained activity after four batches	[97]
9	*Vibrio fluvialis* JS17	Covalent binding	Chitosan	Significant improvement in stability over broad range of pH and temperature	[99]
10	*Vibrio fluvialis*	Covalent binding	Functionalized cellulose	The immobilized enzyme retained activity for four cycles	[100]
11	*Bradyrhizobium japonicum*	Affinity	Ni^{2+}-functionalized polydopamine magneticna nopartices	Simultaneous purification and immobilization of the his-tagged protein could be achieved	[104]
12	*Mycobacterium vanbaalenii*	Covalent binding	magnetic PVA-Fe_3O_4 nanoparticles	The immobilized enzyme could be successfully reused for 13 times in biotransformation	[105]

5. Cascade Reactions Involving ω-TAs

Enzymatic cascades, wherein several independent reactions proceed simultaneously, offer numerous advantages in the production of target compounds, such as shortened synthetic routes, minimization of the use of organic solvents, and improvement in atom efficiency [11,12]. Furthermore, the easier handling of unstable intermediates and the control of unfavorable reaction equilibrium are added benefits of using biocatalytic cascades [107–109]. Cascade reactions have been categorized into four main types: (1) Linear cascades represent consecutive transformations in one-pot. (2) Parallel cascades, which are frequently used in redox biocatalysis wherein the product formation is coupled with a simultaneously proceeding second reaction. (3) The third type, orthogonal cascades, are closely related to parallel cascades and are generally used to shift the reaction equilibrium towards product formation. (4) Cyclic cascades, which are generally used for the stereoinversion processes. Readers are referred to recent reviews [107–113], which have elegantly elaborated on the advancements achieved in the field of cascade reactions.

Multi-enzymatic cascades involving ω-TAs have been successfully used for the amination of the hydroxyl functional groups of alcohols to their corresponding amines. For instance, Fuchs et al. [114] utilized the cascade of galactose oxidase and ω-TAs for the amination of benzylic and cinnamic alcohols. In the first step, alcohol substrates are converted to their corresponding aldehydes in the presence of galactose oxidase. In the subsequent step, the transamination of aldehyde intermediates arecatalyzed by ω-TAs (Figure 6).

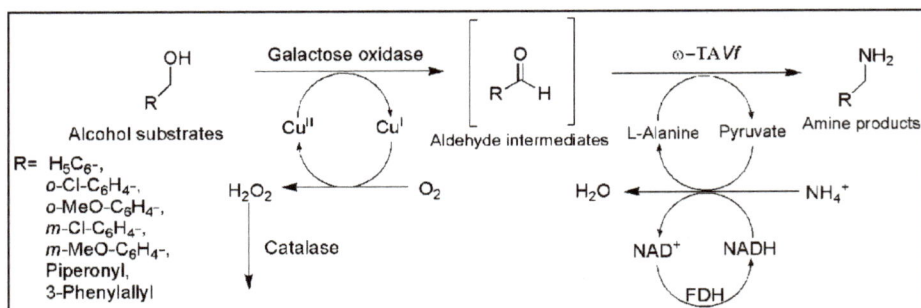

Figure 6. Biocatalytic cascade employing galactose oxidase and ω-TA for the amination of benzylic/cinnamic alcohols (reprinted from [114], with permission of The Royal Society of Chemistry). (Reaction conditions: 50 mM substrate, 20 mg lyophilized *E. coli* BL21(DE3) cells over-expressing galactose oxidase, 0.075 mg/mL 2,2'-azino-bis(3-ethylbenzothiazoline-6-sulfonic acid), 20 mg, lyophilized *E. coli* BL21(DE3) cells over-expressing ω-TA*Vf*, Sodium phosphate buffer (100 mM, pH 7.0), 1 mM PLP, 10 mM CuSO$_4$, 7.5 mg alanine dehydrogenase, 20 U glucose dehydrogenase, 150 mM L-alanine, 120 mM (2.4 equiv) glucose, R.T., 24 h).

Similarly, an artificial redox-neutral cascade, employing alcohol dehydrogenase, alanine dehydrogenase, and ω-TA, has been recently utilized for the conversion of ether alcohol substrates to their corresponding ether amine products [115]. Initially, alcohol ether substrates were converted to the corresponding aldehydes or ketones using ADH from *G. stearothermophilus*. ω-TA*Cv* subsequently catalyzed the amination of these intermediates, wherein AlaDH from *V. proteolyticus* recycled pyruvate to L-alanine with the simultaneous regeneration of the NAD$^+$ cofactor (Figure 7).

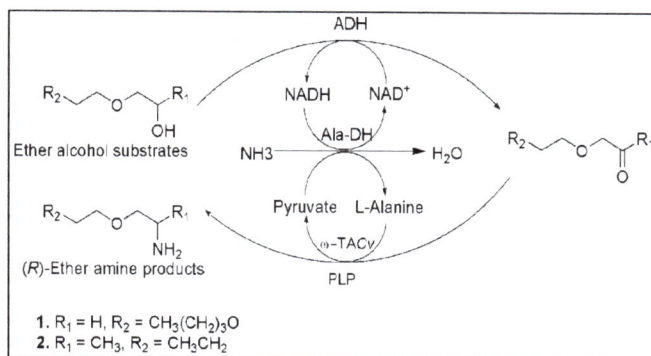

Figure 7. Artificial redox-neutral cascade employing alcohol dehydrogenase, ω-TA and alanine dehydrogenase for the synthesis of ether amines from ether alcohol substrates (reprinted from [115] with the permission of John Wiley and Sons). (Reaction conditions: 10 mM substrate, 280 mM ammonia, ammonium carbamate buffer (40 mM, pH 9.0), 1 mM NAD$^+$, 0.35 mM PLP, 0.5 mM L-Alanine, 0.1 mg/mL ADH, 0.004 mg/mL Ala-DH, 0.064 mg/mL ω-TACv, 30 °C).

Chiral amino-alcohols are known as common constituents of many natural products, pharmaceuticals and chiral auxiliaries used in chemical syntheses [116]. For instance, a chiral amino-triol, (2S,3R)-2-amino-1,3,4-butanetriol (ABT), is used as a precursor for the synthesis of statins and ultimately for a HIV-protease inhibitor Nelfinavir. Gruber et al. [117] recently demonstrated a novel 2-step-enzymatic synthesis of approach, employing transketolase and ω-TA, for the synthesis of ABT in a continuous-flow microreactor system. In the initial step, transketolase -catalyzed asymmetric carbon-carbon bond formation between hydroxypyruvate and glycolaldehyde led to the generation of L-Erythrulose. In the subsequent step, ω-TA-catalyzed conversion of the keto-group into a chiral amino group generated ABT (Figure 8).

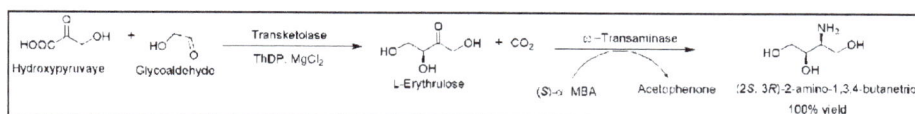

Figure 8. Biocatalytic cascade employing transketolase and ω-TA for the synthesis of amino alcohol (2S,3R)-2-amino-1,3,4-butanetriol (Reprinted from [117], with the permission of John Wiley and Sons). (Reaction conditions: 50 mM substrate, 3.25 U/mL transketolase, 4.8 mM thiamine diphosphate (ThDP), 19.6 mM MgCl$_2$, 10 mM (S)-α-MBA, 27 U/mL ω-TA, 2 mM PLP, Tris-HCl buffer (100 mM, pH 9.0), R.T.).

Despite its complexity, sophisticated enzymatic cascades were recently developed for the synthesis of pharmaceutically important alkaloid scaffolds, such as THIQs. For instance, Erdmann et al. [118] recently reported the synthesis of 1,3,4-trisubstituted THIQs with three chiral centers by employing the cascade of three enzymes, namely carboligase, ω-TACv and a mutant of Pictet-Spenglerase enzyme Norcoclaurine synthase (NCS) from *Thalictrum flavum*. In the first reaction, carboligase catalyzed the generation of (R)-1-hydroxy-1-(3-hydroxyphenyl)propan-2-one from 3-hydroxybenzaldehyde and pyruvate substrates. The transamination catalyzed by ω-TACv generated the corresponding amino product from (R)-1-hydroxy-1-(3-hydroxyphenyl)propan-2-one. Subsequent Pictet-Spengler condensation, catalyzed by NCS-*Tf*, generated a THIQ product (1S,3S,4R)-1-benzyl-3-methyl-1,2,3,4-THIQ-4,6-diol (Figure 9). In addition, opposite stereoselectivities of the C1-substituted-THIQs could be achieved following the chemical cyclization by phosphate (Figure 9), providing access to both the orientations of the substituted THIQ [118].

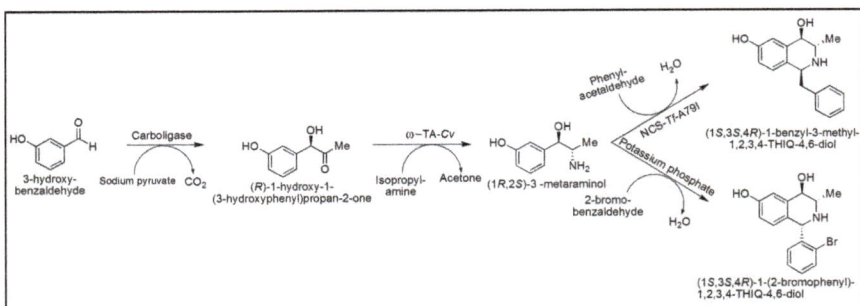

Figure 9. Enzymatic and Chemo-enzymatic synthesis of 1-benzyl-THIQs employing the cascade of carboligase, ω-TA and NCS [118]. (Reaction conditions: step 1: 10 mM substrate, HEPES buffer (100 mM, pH 7.5), 5 mM MgCl$_2$, 0.1 mM ThDP, 0.05 mM FAD, 20 mM sodium pyruvate, 2.5% DMSO (*v/v*), 0.5 mg/mL acetohydroxy acid synthase I from *E. coli* (EcAHAS-I), 30 °C, 750 rpm; Step 2: 0.2 mM PLP, 100 mM isopropylamine, 3.1% DMSO (*v/v*), 3 mg/mL ω-TA*Cv*, 30 °C, 750 rpm; step 3 (Enzymatic): 9.5 mm phenylacetaldehyde, 2.5% DMSO (*v/v*), 0.5 mg/ mL NCS-*Tf*-A79I), 37 °C, 750 rpm; step 3 (Chemo-enzymatic): 10 mM bromobenzaldehyde, 200 mM potassium phosphate buffer, pH 7.0, 50 °C, 750 rpm).

Another Pictet-Spenglerase enzyme, Strictosidine synthase (STR), has also been employed in the cascade consisting of ω-TAs for the synthesis of (*S*)-strictosidine derivatives possessing an additional stereogenic center at C3 of the tetrahydro-β-carboline moiety [119]. In the initial reaction, α-methyltryptamine derivatives were generated using ω-TA-catalyzed amination of prochiral ketone substrates. Subsequent STR-catalyzed condensation of α-methyltryptamine derivatives with secologanin formed optically pure C3-methyl-substituted strictosidine derivatives. Furthermore, the use of stereocomplementary ω-TAs allowed the synthesis of both the enantiomers of α-methyltryptamine derivatives, which ultimately allowed the synthesis of both the epimers of C3-methyl-substituted strictosidine derivatives [119] (Figure 10).

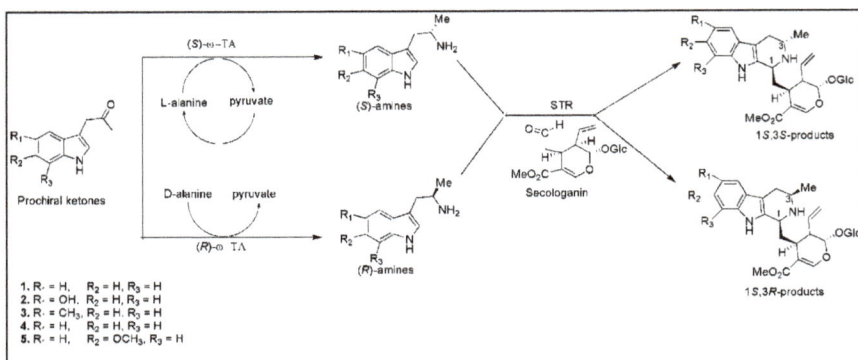

Figure 10. Biocatalytic cascade employing ω-TA and STR for the chemo-enzymatic synthesis of C3-methyl-substituted enantiopure Strictosidine derivatives (adapted with permission from [119], Copyright (2015) American Chemical Society). (Reaction conditions: 2 mM substrate, potassium phosphate buffer (100 mM, pH 7.0), 10 mg lyophilized *E. coli* cells over-expressing ω-TA, 90 mU STR, 250 mM D-or L-alanine, 150 mM ammonium formate, 1 mM PLP, 1 mM NAD$^+$ 11 U FDH, 4 U Ala-DH, DMSO (5%, *v/v*), 4 mM secologanin, 24 h, 30 °C).

6. Conclusions and Prospects

The ever-increasing demands of enzyme-mediated synthesis of enantiopure amines are being elegantly met by engineered ω-TAs. The decreasing costs of gene synthesis and the increasing availability of the genome sequences have given access to a new pool of potential ω-TAs. Therefore, the exploitation of these novel biocatalysts can provide a massive addition to the application of biocatalysts for the industrial synthesis of various pharmaceuticals, fine chemicals, and other important industrial products. This accelerated rate of biocatalyst discovery has permitted scientific community to enter a '*golden age*' of biocatalysis[120]. Despite this progress, many of the recently-discovered enzymes still display poor substrate tolerance profiles, making their role as routine catalysts unlikely. Nevertheless, the recent advances in the reaction engineering such as, development of '*smart*' amine donors, in-situ product removal, and application of cascade reactions, have successfully addressed the major challenge of poor reaction equilibrium towards product formation. In addition, protein engineering strategies have played a pivotal role in expanding the substrate scope of recently discovered biocatalysts and further in improving the thermodynamic and operational stabilities of the already-available enzymes. The future of ω-TA biocatalysis research should also see the discovery of novel enzymes, particularly thermostable ones, by utilizing metagenomic approaches. The application of engineered ω-TAs for the industrial production of sitagliptin has already exhibited the synthetic potential of these potent biocatalysts. Engineered ω-TAs will continue to develop as a competitive alternative technology to conventional syntheses for the production of enantiomerically enriched amines.

Author Contributions: M.D.P. researched the literature and drafted the manuscript; G.G., A.B. and H.Y. discussed ideas and edited the manuscript. All the authors revised and approved the manuscript.

Funding: This research was supported in part by Basic Science Research Program through the National Research Foundation of Korea (NRF) funded by the Ministry of Science, ICT and future Planning (2016R1A2B2014794).

Acknowledgments: M.D.P. gratefully acknowledges 'Foreign Research Professorship' under '2018-KU Brain Pool program' from Konkuk University, Seoul.

Conflicts of Interest: The authors declare no conflict of interest.

References

1. Abaházi, E.; Sátorhelyi, P.; Erdélyi, B.; Vértessy, B.G.; Land, H.; Paizs, C.; Berglund, P.; Poppe, L. Covalently immobilized Trp60Cys mutant of ω-transaminase from *Chromobacterium violaceum* for kinetic resolution of racemic amines in batch and continuous-flow modes. *Biochem. Eng. J.* **2018**, *132*, 270–278. [CrossRef]

2. Zheng, G.W.; Xu, J.H. New opportunities for biocatalysis: Driving the synthesis of chiral chemicals. *Curr. Opin. Biotechnol.* **2011**, *22*, 784–792. [CrossRef] [PubMed]

3. Solano, D.M.; Hoyos, P.; Hernáiz, M.J.; Alcántara, A.R.; Sánchez-Montero, J.M. Industrial biotransformations in the synthesis of building blocks leading to enantiopure drugs. *Bioresour. Technol.* **2012**, *115*, 196–207. [CrossRef] [PubMed]

4. Albarran-Velo, J.; Gonzalez-Martinez, D.; Gotor-Fernandez, V. Stereoselective biocatalysis: A mature technology for the asymmetric synthesis of pharmaceutical building blocks. *Biocatal. Biotransform.* **2018**, *36*, 102–130. [CrossRef]

5. Yang, G.; Ding, Y. Recent advances in biocatalyst discovery, development and applications. *Bioorg. Med. Chem.* **2014**, *22*, 5604–5612. [CrossRef] [PubMed]

6. Simon, R.C.; Mutti, F.G.; Kroutil, W. Biocatalytic synthesis of enantiopure building blocks for pharmaceuticals. *Drug Discov. Today Technol.* **2013**, *10*, e37–e44. [CrossRef] [PubMed]

7. Truppo, M.D. Biocatalysis in the Pharmaceutical Industry—The Need for Speed. *ACS Med. Chem. Lett.* **2017**, *8*, 476–480. [CrossRef] [PubMed]

8. Bornscheuer, U.T. The fourth wave of biocatalysis is approaching. *Phil. Trans. R. Soc. A* **2018**, *376*, 20170063. [CrossRef] [PubMed]

9. Ghislieri, D.; Turner, N.J. Biocatalytic Approaches to the Synthesis of Enantiomerically Pure Chiral Amines. *Top. Catal.* **2014**, *57*, 284–300. [CrossRef]

10. Gomm, A.; O'Reilly, E. Transaminases for chiral amine synthesis. *Curr. Opin. Chem. Biol.* **2018**, *43*, 106–112. [CrossRef] [PubMed]

11. Slabu, I.; Galman, J.L.; Lloyd, R.C.; Turner, N.J. Discovery, Engineering, and Synthetic Application of Transaminase Biocatalysts. *ACS Catal.* **2017**, *7*, 8263–8284. [CrossRef]

12. Guo, F.; Berglund, P. Transaminase biocatalysis: Optimization and application. *Green Chem.* **2017**, *19*, 333–360. [CrossRef]

13. Höhne, M.; Bornscheuer, U.T. Biocatalytic routes to optically active amines. *ChemCatChem* **2009**, *1*, 42–51. [CrossRef]

14. Koszelewski, D.; Tauber, K.; Faber, K.; Kroutil, W. ω-Transaminases for the synthesis of non-racemic α-chiral primary amines. *Trends Biotechnol.* **2010**, *28*, 324–332. [CrossRef] [PubMed]

15. Brundiek, H.; Höhne, M. Transaminases—A Biosynthetic Route for Chiral Amines. In *Applied Biocatalysis: from Fundamental Science to Industrial Applications*; Hilterhaus, L., Liese, A., Kettling, U., Antranikian, G., Eds.; Wiley-VCH Verlag GmbH & Co. KGaA: Weinheim, Germany, 2016; pp. 199–218.

16. Ferrandi, E.E.; Monti, D. Amine transaminases in chiral amines synthesis: Recent advances and challenges. *World J. Microbiol. Biotechnol.* **2018**, *34*, 13. [CrossRef] [PubMed]

17. Busto, E.; Simon, R.C.; Richter, N.; Kroutil, W. Enzymatic Synthesis of Chiral Amines using ω-Transaminases, Amine Oxidases, and the Berberine Bridge Enzyme. In *Green Biocatalysis*; Patel, R.N., Ed.; John Wiley & Sons, Inc.: Hoboken, NJ, USA, 2016; pp. 17–57.

18. Mathew, S.; Yun, H. ω-Transaminases for the production of optically pure amines and unnatural amino acids. *ACS Catal.* **2012**, *2*, 993–1001. [CrossRef]

19. Szmejda, K.; Florczak, T.; Jodłowska, I.; Turkiewicz, M. Extremophilic and modified aminotransferases as a versatile tool for the synthesis of optically pure building blocks for pharmaceutical industry. *Biotechnol. Food Sci.* **2017**, *81*, 23–34.

20. Kelly, S.A.; Pohle, S.; Wharry, S.; Mix, S.; Allen, C.C.; Moody, T.S.; Gilmore, B.F. Application of ω-Transaminases in the Pharmaceutical Industry. *Chem. Rev.* **2018**, *118*, 349–367. [CrossRef] [PubMed]

21. Berglund, P.; Humble, M.S.; Branneby, C. *Comprehensive Chirality*; Carreira, E.M., Yamamoto, H., Eds.; Elsevier: Amsterdam, The Nertherland, 2012; Volume 7, pp. 390–401.

22. Kohls, H.; Steffen-Munsberg, F.; Höhne, M. Recent achievements in developing the biocatalytic toolbox for chiral amine synthesis. *Curr. Opin. Chem. Biol.* **2014**, *19*, 180–192. [CrossRef] [PubMed]

23. Matcham, G.W.; Stirling, D.I.; Zeitlin, A.L. Enantiomeric Enrichment and Stereoselective Synthesis of Chiral amines. U.S. Patent 4950606 A, 5 April 1994.

24. Wu, H.L.; Zhang, J.D.; Zhang, C.F.; Fan, X.J.; Chang, H.H.; Wei, W.L. Characterization of Four New Distinct ω-Transaminases from *Pseudomonas putida* NBRC 14164 for Kinetic Resolution of Racemic Amines and Amino Alcohols. *Appl. Biochem. Biotechnol.* **2017**, *181*, 972–985. [CrossRef] [PubMed]

25. Höhne, M.; Robins, K.; Bornscheuer, U.T. A Protection Strategy Substantially Enhances Rate and Enantioselectivity in ω-Transaminase-Catalyzed Kinetic Resolutions. *Adv. Synth. Catal.* **2008**, *350*, 807–812. [CrossRef]

26. Kaulmann, U.; Smithies, K.; Smith, M.E.; Hailes, H.C.; Ward, J.M. Substrate spectrum of ω-transaminase from *Chromobacterium violaceum* DSM30191 and its potential for biocatalysis. *Enzyme Microb. Technol.* **2007**, *41*, 628–637. [CrossRef]

27. Hanson, R.L.; Davis, B.L.; Chen, Y.; Goldberg, S.L.; Parker, W.L.; Tully, T.P.; Montana, M.A.; Patel, R.N. Preparation of (*R*)-Amines from Racemic Amines with an (*S*)-Amine Transaminase from *Bacillus megaterium*. *Adv. Synth. Catal.* **2008**, *350*, 1367–1375. [CrossRef]

28. Malik, M.S.; Park, E.S.; Shin, J.S. ω-Transaminase-catalyzed kinetic resolution of chiral amines using L-threonine as an amino acceptor precursor. *Green Chem.* **2012**, *14*, 2137–2140. [CrossRef]

29. Savile, C.K.; Janey, J.M.; Mundorff, E.C.; Moore, J.C.; Tam, S.; Jarvis, W.R.; Colbeck, J.C.; Krebber, A.; Fleitz, F.J.; Brands, J.; et al. Biocatalytic asymmetric synthesis of chiral amines from ketones applied to Sitagliptin manufacture. *Science* **2010**, *329*, 305–309. [CrossRef] [PubMed]

30. Gomm, A.; Lewis, W.; Green, A.P.; O'Reilly, E. A New Generation of Smart Amine Donors for Transaminase-Mediated Biotransformations. *Chem. Eur. J.* **2016**, *22*, 12692–12695. [CrossRef] [PubMed]

31. Stekhanova, T.N.; Rakitin, A.L.; Mardanov, A.V.; Bezsudnova, E.Y.; Popov, V.O. A Novel highly thermostable branched-chain amino acid aminotransferase from the crenarchaeon *Vulcanisaeta moutnovskia*. *Enzyme Microb. Technol.* **2017**, *96*, 127–134. [CrossRef] [PubMed]

32. Sayer, C.; Martinez-Torres, R.J.; Richter, N.; Isupov, M.N.; Hailes, H.C.; Littlechild, J.A.; Ward, J.M. The substrate specificity, enantioselectivity and structure of the (*R*)-selective amine: Pyruvate transaminase from *Nectria haematococca*. *FEBS J.* **2014**, *281*, 2240–2253. [CrossRef] [PubMed]

33. Mathew, S.; Nadarajan, S.P.; Chung, T.; Park, H.H.; Yun, H. Biochemical characterization of thermostable ω-transaminase from *Sphaerobacter thermophilus* and its application for producing aromatic β-and γ-amino acids. *Enzyme Microb. Technol.* **2016**, *87*, 52–60. [CrossRef] [PubMed]

34. Iglesias, C.; Panizza, P.; Giordano, S.R. Identification, expression and characterization of an *R*-ω-transaminase from *Caproniasemi immersa*. *Appl. Microbiol. Biotechnol.* **2017**, *101*, 5677–5687. [CrossRef] [PubMed]

35. Baud, D.; Jeffries, J.W.; Moody, T.S.; Ward, J.M.; Hailes, H.C. A metagenomics approach for new biocatalyst discovery: Application to transaminases and the synthesis of allylic amines. *Green Chem.* **2017**, *19*, 1134–1143. [CrossRef]

36. Gao, S.; Su, Y.; Zhao, L.; Li, G.; Zheng, G. Characterization of a (*R*)-selective amine transaminase from *Fusarium oxysporum*. *Process Biochem.* **2017**, *63*, 130–136. [CrossRef]

37. Yun, H.; Cho, B.K.; Kim, B.G. Kinetic resolution of (*R, S*)-sec-butylamine using omega-transaminase from *Vibrio fluvialis* JS17 under reduced pressure. *Biotechnol. Bioeng.* **2004**, *87*, 772–778. [CrossRef] [PubMed]

38. Tufvesson, P.; Bach, C.; Woodley, J.M. A model to assess the feasibility of shifting reaction equilibrium by acetone removal in the transamination of ketones using 2-propylamine. *Biotechnol. Bioeng.* **2014**, *111*, 309–319. [CrossRef] [PubMed]

39. Hou, A.; Deng, Z.; Ma, H.; Liu, T. Substrate screening of amino transaminase for the synthesis of a sitagliptin intermediate. *Tetrahedron* **2016**, *72*, 4660–4664. [CrossRef]

40. Park, E.S.; Shin, J.S. Biocatalytic cascade reactions for asymmetric synthesis of aliphatic amino acids in a biphasic reaction system. *J. Mol. Catal. B Enzym.* **2015**, *121*, 9–14. [CrossRef]

41. Neto, W.; Schurmann, M.; Panella, L.; Vogel, A.; Woodley, J.M. Immobilization of-transaminase for industrial application: Screening and characterization of commercial ready to use enzyme carriers. *J. Mol. Catal. B Enzym.* **2015**, *117*, 54–61. [CrossRef]

42. Rehn, G.; Grey, C.; Branneby, C.; Lindberg, L.; Adlercreutz, P. Activity and stability of different immobilized preparations of recombinant *E. coli* cells containing ω-transaminase. *Process Biochem.* **2012**, *47*, 1129–1134. [CrossRef]

43. Bajić, M.; Plazl, I.; Stloukal, R.; Žnidaršič-Plazl, P. Development of a miniaturized packed bed reactor with ω-transaminase immobilized in LentiKats®. *Process Biochem.* **2017**, *52*, 63–72. [CrossRef]

44. Satyawali, Y.; Ehimen, E.; Cauwenberghs, L.; Maesen, M.; Vandezande, P.; Dejonghe, W. Asymmetric synthesis of chiral amine in organic solvent and in-situ product recovery for process intensification: A case study. *Biochem. Eng. J.* **2017**, *117*, 97–104. [CrossRef]

45. Lalonde, J. Highly engineered biocatalysts for efficient small molecule pharmaceutical synthesis. *Curr. Opin. Biotechnol.* **2016**, *42*, 152–158. [CrossRef] [PubMed]

46. López-Iglesias, M.; González-Martínez, D.; Gotor, V.; Busto, E.; Kroutil, W.; Gotor-Fernández, V. Biocatalytic Transamination for the Asymmetric Synthesis of Pyridylalkylamines. Structural and Activity Features in the Reactivity of Transaminases. *ACS Catal.* **2016**, *6*, 4003–4009. [CrossRef]

47. Busto, E.; Simon, R.C.; Grischek, B.; Gotor-Fernández, V.; Kroutil, W. Cutting short the asymmetric synthesis of the Ramatroban precursor by employing ω-Transaminases. *Adv. Synth. Catal.* **2014**, *356*, 1937–1942. [CrossRef]

48. Costa, I.C.; De Souza, R.O.; Bornscheuer, U.T. Asymmetric synthesis of serinol-monoesters catalyzed by amine transaminases. *Tetrahedron Asymmetry* **2017**, *28*, 1183–1187. [CrossRef]

49. Martin, A.R.; Shonnard, D.; Pannuri, S.; Kamat, S. Characterization of a High Activity (*S*)-Aminotransferase for Substituted (*S*)-Aminotetralin Production: Properties and Kinetics. *J. Bioprocess. Biotech.* **2011**, *1*, 107. [CrossRef]

50. Ito, N.; Kawano, S.; Hasegawa, J.; Yasohara, Y. Purification and Characterization of a Novel (*S*)-Enantioselective Transaminase from *Pseudomonas fluorescens* KNK08-18 for the Synthesis of Optically Active Amines. *Biosci. Biotechnol. Biochem.* **2011**, *75*, 2093–2098. [CrossRef] [PubMed]

51. Voges, M.; Schmidt, F.; Wolff, D.; Sadowski, G.; Held, C. Thermodynamics of the alanine aminotransferase reaction. *Fluid Ph. Equilibria* **2016**, *422*, 87–98. [CrossRef]

52. Seo, J.H.; Kyung, D.; Joo, K.; Lee, J.; Kim, B.G. Necessary and sufficient conditions for the asymmetric synthesis of chiral amines using ω-aminotransferases. *Biotechnol. Bioeng.* **2011**, *108*, 253–263. [CrossRef] [PubMed]

53. Green, A.P.; Turner, N.J.; O'Reilly, E. Chiral amine synthesis using ω-transaminases: An amine donor that displaces equilibria and enables high-throughput screening. *Angew. Chem. Int. Ed.* **2014**, *53*, 10714–10717. [CrossRef] [PubMed]

54. Richter, N.; Farnberger, J.E.; Pressnitz, D.; Lechner, H.; Zepeck, F.; Kroutil, W. A system for ω-transaminase mediated (*R*)-amination using L-alanine as an amine donor. *Green Chem.* **2015**, *17*, 2952–2958. [CrossRef]

55. Martínez-Montero, L.; Gotor, V.; Gotor-Fernández, V.; Lavandera, I. Stereoselective amination of racemic sec-alcohols through sequential application of laccases and transaminases. *Green Chem.* **2017**, *19*, 474–480. [CrossRef]

56. Galman, J.L.; Slabu, I.; Weise, N.J.; Iglesias, C.; Parmeggiani, F.; Lloyd, R.C.; Turner, N.J. Biocatalytic transamination with near-stoichiometric inexpensive amine donors mediated by bifunctional mono-and di-amine transaminases. *Green Chem.* **2017**, *19*, 361–366. [CrossRef]

57. Mathew, S.; Nadarajan, S.P.; Sundaramoorthy, U.; Jeon, H.; Chung, T.; Yun, H. Biotransformation of β-keto nitriles to chiral (*S*)-β-amino acids using nitrilase and ω-transaminase. *Biotechnol. Lett.* **2017**, *39*, 535–543. [CrossRef] [PubMed]

58. Rehn, G.; Ayres, B.; Adlercreutz, P.; Grey, C. An improved process for biocatalytic asymmetric amine synthesis by in situ product removal using a supported liquid membrane. *J. Mol. Catal. B Enzym.* **2016**, *123*, 1–7. [CrossRef]

59. Rudroff, F.; Mihovilovic, M.D.; Gröger, H.; Snajdrova, R.; Iding, H.; Bornscheuer, U.T. Opportunities and challenges for combining chemo-and biocatalysis. *Nat. Catal.* **2018**, *1*, 12–22. [CrossRef]

60. Kroutil, W.; Fischereder, E.M.; Fuchs, C.S.; Lechner, H.; Mutti, F.G.; Pressnitz, D.; Rajagopalan, A.; Sattler, J.H.; Simon, R.C.; Siirola, E. Asymmetric preparation of prim, sec-, and tert-amines employing selected biocatalysts. *Org. Proc. Res. Dev.* **2013**, *17*, 751–759. [CrossRef] [PubMed]

61. Koszelewski, D.; Pressnitz, D.; Clay, D.; Kroutil, W. Deracemization of mexiletine biocatalyzed by ω-transaminases. *Org. Lett.* **2009**, *11*, 4810–4812. [CrossRef] [PubMed]

62. Chung, C.K.; Bulger, P.G.; Kosjek, B.; Belyk, K.M.; Rivera, N.; Scott, M.E.; Humphrey, G.R.; Limanto, J.; Bachert, D.C.; Emerson, K.M. Process development of C–N cross-coupling and enantioselective biocatalytic reactions for the asymmetric synthesis of niraparib. *Org. Proc. Res. Dev.* **2013**, *18*, 215–227. [CrossRef]

63. Peng, Z.; Wong, J.W.; Hansen, E.C.; Puchlopek-Dermenci, A.L.; Clarke, H.J. Development of a concise, asymmetric synthesis of a smoothened receptor (SMO) inhibitor: Enzymatic transamination of a 4-piperidinone with dynamic kinetic resolution. *Org. Lett.* **2014**, *16*, 860–863. [CrossRef] [PubMed]

64. Koszelewski, D.; Clay, D.; Rozzell, D.; Kroutil, W. Deracemisation of α-chiral primary amines by a one-pot, two-step cascade reaction catalysed by ω-transaminases. *Eur. J. Org. Chem.* **2009**, *14*, 2289–2292. [CrossRef]

65. Shin, G.; Mathew, S.; Shon, M.; Kim, B.G.; Yun, H. One-pot one-step deracemization of amines using ω-transaminases. *Chem. Commun.* **2013**, *49*, 8629–8631. [CrossRef] [PubMed]

66. Steffen-Munsberg, F.; Vickers, C.; Thontowi, A.; Schätzle, S.; Tumlirsch, T.; Svedendahl-Humble, M.; Land, H.; Berglund, P.; Bornscheuer, U.T.; Höhne, M. Connecting unexplored protein crystal structures to enzymatic function. *ChemCatChem* **2013**, *5*, 150–153. [CrossRef]

67. Shin, J.S.; Kim, B.G. Exploring the Active Site of Amine: Pyruvate Aminotransferase on the Basis of the Substrate Structure-Reactivity Relationship: How the Enzyme Controls Substrate Specificity and Stereoselectivity. *J. Org. Chem.* **2002**, *67*, 2848–2853. [CrossRef] [PubMed]

68. Nobili, A.; Steffen-Munsberg, F.; Kohls, H.; Trentin, I.; Schulzke, C.; Höhne, M.; Bornscheuer, U.T. Engineering the active site of the amine transaminase from *Vibrio fluvialis* for the asymmetric synthesis of aryl–alkyl amines and amino alcohols. *ChemCatChem* **2015**, *7*, 757–760. [CrossRef]

69. Steffen-Munsberg, F.; Vickers, C.; Thontowi, A.; Schätzle, S.; Meinhardt, T.; Svedendahl-Humble, M.; Land, H.; Berglund, P.; Bornscheuer, U.T.; Höhne, M. Revealing the structural basis of promiscuous amine transaminase activity. *ChemCatChem* **2013**, *5*, 154–157. [CrossRef]

70. Desai, A.A. Sitagliptin manufacture: A compelling tale of green chemistry, process intensification, and industrial asymmetric catalysis. *Angew. Chem. Int. Ed.* **2011**, *50*, 1974–1976. [CrossRef] [PubMed]

71. Midelfort, K.S.; Kumar, R.; Han, S.; Karmilowicz, M.J.; McConnell, K.; Gehlhaar, D.K.; Mistry, A.; Chang, J.S.; Anderson, M.; Villalobos, A.; et al. Redesigning and characterizing the substrate specificity and activity of *Vibrio fluvialis* aminotransferase for the synthesis of Imagabalin. *Protein Eng. Des. Sel.* **2012**, *26*, 25–33. [CrossRef] [PubMed]

72. Genz, M.; Melse, O.; Schmidt, S.; Vickers, C.; Dörr, M.; Van den Bergh, T.; Joosten, H.J.; Bornscheuer, U.T. Engineering the Amine Transaminase from *Vibrio fluvialis* towards Branched-Chain Substrates. *ChemCatChem* **2016**, *8*, 3199–3202. [CrossRef]

73. Dourado, D.F.; Pohle, S.; Carvalho, A.T.; Dheeman, D.S.; Caswell, J.M.; Skvortsov, T.; Miskelly, I.; Brown, R.T.; Quinn, D.J.; Allen, C.C.; et al. Rational Design of a (*S*)-Selective-Transaminase for Asymmetric Synthesis of (1*S*)-1-(1, 1′-biphenyl-2-yl) ethanamine. *ACS Catal.* **2016**, *6*, 7749–7759. [CrossRef]

74. Han, S.W.; Park, E.S.; Dong, J.Y.; Shin, J.S. Active-Site Engineering of ω-Transaminase for Production of Unnatural Amino Acids Carrying a Side Chain Bulkier than an Ethyl Substituent. *Appl. Environ. Microbiol.* **2015**, *81*, 6994–7002. [CrossRef] [PubMed]

75. Pavlidis, I.V.; Weiß, M.S.; Genz, M.; Spurr, P.; Hanlon, S.P.; Wirz, B.; Iding, H.; Bornscheuer, U.T. Identification of (*S*)-selective transaminases for the asymmetric synthesis of bulky chiral amines. *Nat. Chem.* **2016**, *8*, 1076–1082. [CrossRef] [PubMed]

76. Weiß, M.S.; Pavlidis, I.V.; Spurr, P.; Hanlon, S.P.; Wirz, B.; Iding, H.; Bornscheuer, U.T. Protein-engineering of an amine transaminase for the stereoselective synthesis of a pharmaceutically relevant bicyclic amine. *Org. Biomol. Chem.* **2016**, *14*, 10249–10254. [CrossRef] [PubMed]

77. Weiß, M.S.; Pavlidis, I.V.; Spurr, P.; Hanlon, S.P.; Wirz, B.; Iding, H.; Bornscheuer, U.T. Amine Transaminase Engineering for Spatially Bulky Substrate Acceptance. *ChemBioChem* **2017**, *18*, 1022–1026. [CrossRef] [PubMed]

78. Polizzi, K.M.; Bommarius, A.S.; Broering, J.M.; Chaparro-Riggers, J.F. Stability of biocatalysts. *Curr. Opin. Chem. Biol.* **2007**, *11*, 220–225. [CrossRef] [PubMed]

79. Stepankova, V.; Bidmanova, S.; Koudelakova, T.; Prokop, Z.; Chaloupkova, R.; Damborsky, J. Strategies for stabilization of enzymes in organic solvents. *ACS Catal.* **2013**, *3*, 2823–2836. [CrossRef]

80. Littlechild, J.A. Enzymes from extreme environments and their industrial applications. *Front. Bioeng. Biotechnol.* **2015**, *3*, 161. [CrossRef] [PubMed]

81. Cerioli, L.; Planchestainer, M.; Cassidy, J.; Tessaro, D.; Paradisi, F. Characterization of a novel amine transaminase from *Halomonas elongata*. *J. Mol. Catal. B Enzym.* **2015**, *120*, 141–150. [CrossRef]

82. Chen, S.; Land, H.; Berglund, P.; Humble, M.S. Stabilization of an amine transaminase for biocatalysis. *J. Mol. Catal. B Enzym.* **2016**, *124*, 20–28. [CrossRef]

83. Mathew, S.; Deepankumar, K.; Shin, G.; Hong, E.Y.; Kim, B.G.; Chung, T.; Yun, H. Identification of novel thermostable ω-transaminase and its application for enzymatic synthesis of chiral amines at high temperature. *RSC Adv.* **2016**, *6*, 69257–69260. [CrossRef]

84. Chen, Y.; Yi, D.; Jiang, S.; Wei, D. Identification of novel thermostable taurine–pyruvate transaminase from *Geobacillus thermodenitrificans* for chiral amine synthesis. *Appl. Microbiol. Biotechnol.* **2016**, *100*, 3101–3111. [CrossRef] [PubMed]

85. Ferrandi, E.E.; Previdi, A.; Bassanini, I.; Riva, S.; Peng, X.; Monti, D. Novel thermostable amine transferases from hot spring metagenomes. *Appl. Microbiol. Biotechnol.* **2017**, *101*, 4963–4979. [CrossRef] [PubMed]

86. Pannuri, S.; Kamat, S.; Venkatesh; Garcia, A.; Rogelio, M. Thermostable Omega-Transaminases. PCT Application No. WO/2006/063336.15 June 2006.

87. Sayer, C.; Isupov, M.N.; Westlake, A.; Littlechild, J.A. Structural studies of *Pseudomonas* and *Chromobacterium* ω-aminotransferases provide insights into their differing substrate specificity. *Acta Crystallogr. D Biol. Crystallogr.* **2013**, *69*, 564–576. [CrossRef] [PubMed]

88. Börner, T.; Rämisch, S.; Reddem, E.R.; Bartsch, S.; Vogel, A.; Thunnissen, A.M.; Adlercreutz, P.; Grey, C. Explaining Operational Instability of Amine Transaminases: Substrate-Induced Inactivation Mechanism and Influence of Quaternary Structure on Enzyme–Cofactor Intermediate Stability. *ACS Catal.* **2017**, *7*, 1259–1269. [CrossRef]

89. Börner, T.; Rämisch, S.; Bartsch, S.; Vogel, A.; Adlercreutz, P.; Grey, C. Three In One Go: Thermo-, Solvent and Catalytic Stability by Engineering the Cofactor-Binding Element of Amine Transaminases. *ChemBioChem* **2017**, *18*, 1482–1486. [CrossRef] [PubMed]

90. Huang, J.; Xie, D.F.; Feng, Y. Engineering thermostable (*R*)-selective amine transaminase from *Aspergillus terreus* through in silico design employing B-factor and folding free energy calculations. *Biochem. Biophys. Res. Commun.* **2017**, *483*, 397–402. [CrossRef] [PubMed]

91. Deepankumar, K.; Nadarajan, S.P.; Mathew, S.; Lee, S.G.; Yoo, T.H.; Hong, E.Y.; Kim, B.G.; Yun, H. Engineering Transaminase for Stability Enhancement and Site-Specific Immobilization through Multiple Noncanonical Amino Acids Incorporation. *ChemCatChem* **2015**, *7*, 417–421. [CrossRef]

92. Deepankumar, K.; Shon, M.; Nadarajan, S.P.; Shin, G.; Mathew, S.; Ayyadurai, N.; Kim, B.G.; Choi, S.H.; Lee, S.H.; Yun, H. Enhancing Thermostability and Organic Solvent Tolerance of ω-Transaminase through Global Incorporation of Fluorotyrosine. *Adv. Synth. Catal.* **2014**, *356*, 993–998. [CrossRef]

93. Patil, M.D.; Dev, M.J.; Shinde, A.S.; Bhilare, K.D.; Patel, G.; Chisti, Y.; Banerjee, U.C. Surfactant-mediated permeabilization of *Pseudomonas putida* KT2440 and use of the immobilized permeabilized cells in biotransformation. *Process Biochem.* **2017**, *63*, 113–121. [CrossRef]

94. Yewale, T.; Singhal, R.S.; Vaidya, A.A. Immobilization of inulinase from *Aspergillus niger* NCIM 945 on chitosan and its application in continuous inulin hydrolysis. *Biocatal. Agric. Biotechnol.* **2013**, *2*, 96–101. [CrossRef]

95. Koszelewski, D.; Müller, N.; Schrittwieser, J.H.; Faber, K.; Kroutil, W. Immobilization of ω-transaminases by encapsulation in a sol–gel/celite matrix. *J. Mol. Catal. B Enzym.* **2010**, *63*, 39–44. [CrossRef]

96. Mallin, H.; Menyes, U.; Vorhaben, T.; Höhne, M.; Bornscheuer, U.T. Immobilization of two (*R*)-Amine Transaminases on an Optimized Chitosan Support for the Enzymatic Synthesis of Optically Pure Amines. *ChemCatChem* **2013**, *5*, 588–593. [CrossRef]

97. Mallin, H.; Höhne, M.; Bornscheuer, U.T. Immobilization of (*R*)-and (*S*)-amine transaminases on chitosan support and their application for amine synthesis using isopropylamine as donor. *J. Biotechnol.* **2014**, *191*, 32–37. [CrossRef] [PubMed]

98. Truppo, M.D.; Strotman, H.; Hughes, G. Development of an immobilized transaminase capable of operating in organic solvent. *ChemCatChem* **2012**, *4*, 1071–1074. [CrossRef]

99. Yi, S.S.; Lee, C.W.; Kim, J.; Kyung, D.; Kim, B.G.; Lee, Y.S. Covalent immobilization of ω-transaminase from *Vibrio fluvialis* JS17 on chitosan beads. *Process Biochem.* **2007**, *42*, 895–898. [CrossRef]

100. De Souza, S.P., II; Silva, G.M.; Miranda, L.S.; Santiago, M.F.; Lam, F.L.; Dawood, A.; Bornscheuer, U.T.; De Souza, R.O. Cellulose as an efficient matrix for lipase and transaminase immobilization. *RSC Adv.* **2016**, *6*, 6665–6671. [CrossRef]

101. Cárdenas-Fernández, M.; Neto, W.; López, C.; Alvaro, G.; Tufvesson, P.; Woodley, J.M. Immobilization of *Escherichia coli* containing ω-transaminase activity in LentiKats®. *Biotechnol. Prog.* **2012**, *28*, 693–698. [CrossRef] [PubMed]

102. Matosevic, S.; Lye, G.J.; Baganz, F. Immobilised enzyme microreactor for screening of multi-step bioconversions: Characterisation of a de novo transketolase-ω-transaminase pathway to synthesise chiral amino alcohols. *J. Biotechnol.* **2011**, *155*, 320–329. [CrossRef] [PubMed]

103. Halim, A.A.; Szita, N.; Baganz, F. Characterization and multi-step transketolase-ω-transaminase bioconversions in an immobilized enzyme microreactor (IEMR) with packed tube. *J. Biotechnol.* **2013**, *168*, 567–575. [CrossRef] [PubMed]

104. Yang, J.; Ni, K.; Wei, D.; Ren, Y. One-step purification and immobilization of his-tagged protein via Ni^{2+}-functionalized Fe_3O_4@ polydopamine magnetic nanoparticles. *Biotechnol. Bioprocess Eng.* **2015**, *20*, 901–907. [CrossRef]

105. Jia, H.; Huang, F.; Gao, Z.; Zhong, C.; Zhou, H.; Jiang, M.; Wei, P. Immobilization of ω-transaminase by magnetic PVA-Fe_3O_4 nanoparticles. *Biotechnol. Rep.* **2016**, *10*, 49–55. [CrossRef] [PubMed]

106. Sun, J.; Cui, W.H.; Du, K.; Gao, Q.; Du, M.; Ji, P.; Feng, W. Immobilization of *R*-ω-transaminase on MnO_2 nanorods for catalyzing the conversion of (*R*)-1-phenylethylamine. *J. Biotechnol.* **2017**, *245*, 14–20. [CrossRef] [PubMed]

107. Sperl, J.M.; Sieber, V. Multienzyme Cascade Reactions- Status and Recent Advances. *ACS Catal.* **2018**, *8*, 2385–2396. [CrossRef]

108. Ahsan, M.M.; Jeon, H.; Nadarajan, S.P.; Chung, T.; Yoo, H.W.; Kim, B.G.; Patil, M.D.; Yun, H. Biosynthesis of the Nylon 12 Monomer, ω-Aminododecanoic Acid with Novel CYP153A, AlkJ, and ω-TA Enzymes. *Biotechnol. J.* **2018**, *13*, 1700562. [CrossRef] [PubMed]

109. Ahsan, M.M.; Sung, S.; Jeon, H.; Patil, M.D.; Chung T Yun, H. Biosynthesis of Medium-to Long-Chain α, ω-Diols from Free Fatty Acids Using CYP153A Monooxygenase, Carboxylic Acid Reductase, and *E. coli* Endogenous Aldehyde Reductases. *Catalysts* **2017**, *8*, 4. [CrossRef]
110. Simon, R.C.; Richter, N.; Busto, E.; Kroutil, W. Recent developments of cascade reactions involving ω-transaminases. *ACS Catal.* **2013**, *4*, 129–143. [CrossRef]
111. Hepworth, L.J.; France, S.P.; Hussain, S.; Both, P.; Turner, N.J.; Flitsch, S.L. Enzyme cascades in whole cells for the synthesis of chiral cyclic amines. *ACS Catal.* **2017**, *7*, 2920–2925. [CrossRef]
112. Fuchs, M.; Farnberger, J.E.; Kroutil, W. The industrial age of biocatalytic transamination. *Eur. J. Org. Chem.* **2015**, *32*, 6965–6982. [CrossRef] [PubMed]
113. France, S.P.; Hepworth, L.J.; Turner, N.J.; Flitsch, S.L. Constructing biocatalytic cascades: In vitro and in vivo approaches to de novo multi-enzyme pathways. *ACS Catal.* **2016**, *7*, 710–724. [CrossRef]
114. Fuchs, M.; Tauber, K.; Sattler, J.; Lechner, H.; Pfeffer, J.; Kroutil, W.; Faber, K. Amination of benzylic and cinnamic alcohols via a biocatalytic, aerobic, oxidation–transamination cascade. *RSC Adv.* **2012**, *2*, 6262–6265. [CrossRef]
115. Palacio, C.M.; Crismaru, C.G.; Bartsch, S.; Navickas, V.; Ditrich, K.; Breuer, M.; Abu, R.; Woodley, J.M.; Baldenius, K.; Wu, B.; et al. Enzymatic network for production of ether amines from alcohols. *Biotechnol. Bioeng.* **2016**, *113*, 1853–1861. [CrossRef] [PubMed]
116. Birrell, J.A.; Jacobsen, E.N. A Practical method for the synthesis of highly enantioenriched trans-1, 2-amino alcohols. *Org. Lett.* **2013**, *15*, 2895–2897. [CrossRef] [PubMed]
117. Gruber, P.; Carvalho, F.; Marques, M.P.; O'Sullivan, B.; Subrizi, F.; Dobrijevic, D.; Ward, J.; Hailes, H.C.; Fernandes, P.; Wohlgemuth, R.; et al. Enzymatic synthesis of chiral amino-alcohols by coupling transketolase and transaminase-catalyzed reactions in a cascading continuous-flow microreactor system. *Biotechnol. Bioeng.* **2018**, *115*, 586–596. [CrossRef] [PubMed]
118. Erdmann, V.; Lichman, B.R.; Zhao, J.; Simon, R.C.; Kroutil, W.; Ward, J.M.; Hailes, H.C.; Rother, D. Enzymatic and Chemoenzymatic Three-Step Cascades for the Synthesis of Stereochemically Complementary Trisubstituted Tetrahydroisoquinolines. *Angew. Chem. Int. Ed.* **2017**, *56*, 12503–12507. [CrossRef] [PubMed]
119. Fischereder, E.M.; Pressnitz, D.; Kroutil, W. Stereoselective Cascade to C3-methylated strictosidine derivatives employing transaminases and strictosidine synthases. *ACS Catal.* **2015**, *6*, 23–30. [CrossRef]
120. Turner, N.J.; Kumar, R. Editorial overview: Biocatalysis and biotransformation: The golden age of biocatalysis. *Curr. Opin. Chem. Biol.* **2018**, *43*, A1–A3. [CrossRef] [PubMed]

MDPI

St. Alban-Anlage 66

4052 Basel

Switzerland

Tel. +41 61 683 77 34

Fax +41 61 302 89 18

www.mdpi.com

Catalysts Editorial Office

E-mail: catalysts@mdpi.com

www.mdpi.com/journal/catalysts

www.ingramcontent.com/pod-product-compliance
Lightning Source LLC
Chambersburg PA
CBHW051914210326
41597CB00033B/6138